DINOSAUR!

Dr David Norman is Director of the Sedgwick Museum of Geology at the University of Cambridge and he lectures to undergraduates in the departments of Earth Sciences and Zoology on vertebrate evolution and paleobiology. As a research scientist, he has published many academic papers on fossil reptiles, especially dinosaurs.

Dr Norman is also interested in promoting the popular understanding of science: he is well known as the presenter and scientific consultant to several television series, has given many public lectures and published a number of award-winning books explaining the scientific background to the study of dinosaurs. Some of his books, such as the *Illustrated Encyclopaedia of Dinosaurs* (1985) have received wide critical acclaim among both scientists and the general public, and his book *Dinosaur* (1989) written jointly with Angela Milner was shortlisted for Science Book of the Year in 1990. In 1992 the international bestseller *Dinosaur!* (first edition 1991), published by Boxtree, was commended as Family Book of the Year for the Science Book of the Year awards, and was selected by *Library Journal* as one of the best science books of the year for general readers.

One of the world's leading authorities on dinosaurs, Dr Norman is currently collaborating with colleagues in Russia, North America, Romania and South Africa, as well as studying early British dinosaurs. In recent years he has been on a expedition to southern Africa looking for very early dinosaurs, and is developing links with research scientists in South America.

DINOSAUR!

Dr David Norman

Macmillan • USA

Dedication

If they will permit me, I would like to dedicate this book to two living scientists,
Emeritus Professor Edwin H. Colbert and Professor John H. Ostrom who between
them have done much to further the understanding which we now have of
dinosaurs. I would also like to offer this as a tribute to the memory of Gideon
Algernon Mantell FRS (1790–1852), the discoverer and describer of an animal that
is very close to my own heart – *Iguanodon* – one of the first dinosaurs to be found.
My only regret in making this gesture is that I will never know whether Dr Mantell
would have approved.

MACMILLAN
A Simon & Schuster Macmillan Company
1633 Broadway
New York, NY 10019

Copyright © (text) 1991 by David Norman

Library of Congress Cataloging-in-Publication Data

Norman, David, 1930–
    Dinosaur! / David Norman.
        p.   cm.
    Based on the four-part television series with the same title.
    Includes bibliographical references (p.   –   ) and index.
    ISBN 0-02-860434-2 (pbk.)
    1. Dinosaurs.   2. Palaontology—Mesozoic.   I. Dinosaur!
(Television program)   II. Title.
QE862.D5N669   1995
567.9′ 1—dc20   95–19199   CIP

Designed by Robert Updegraft
Typeset by Wilcom Services, London
Origination by Pica Colour Overseas, Singapore
Printed and bound in Italy by New Interlitho, Milan

10 9 8 7 6 5 4 3 2 1

First published in the UK in 1991 by BOXTREE LIMITED Broadwall
House London SE1 9PL

Based on the television series *Dinosaur!* a co-production by Granada
Television (United Kingdom)
Satel (Austria) and Primedia (Canada), in association with Arts &
Entertainment (USA) and ORF (Austria).

Contents

PREFACE

Dinosaurs fascinate children and adults alike, and probably always will. That is hardly surprising. Many are startlingly large creatures which bring a frightening reality to the myth of the dragon – the beast of dread which haunts our subconscious imagination. Childhood is the time of unbounded imagination; none of the rules about what is real and what is not real, what is possible and what is impossible, have yet been firmly fixed in us. To a child, a dinosaur skeleton in a museum can be almost as awe-inspiring as the real, bloodcurdling beast.

In recent years the impact of dinosaurs upon both young children and adults has been dramatically enhanced by sophisticated computer-enhanced graphics and models in films, such as the "blockbuster" *Jurassic Park*, as well as the very popular moving dinosaur displays (built by Kokoro and Dinamation Corporation) which have been touring museums worldwide. The popularity of dinosaurs seemed to reach an all-time high in 1993; dinosaurs have become media stars in their own right and a part of the corporate publicity strategy of Hollywood film makers. Such attention has been decidedly double-edged, however. On one hand, I am sure that the increasing interest in dinosaurs will have encouraged people to visit museums more than they might have in the past, and perhaps led them to discover that there is more to museums than just dinosaurs. But equally it has been the case that expectations about what paleontologists are doing in their research have been raised to preposterous heights: for example, the re-creation of living dinosaurs as seen in *Jurassic Park* is completely out of the question. It has also led media people and other non-experts to say some very stupid things about dinosaurs, which have either simply revealed their utter ignorance of the subject, or seriously misled people.

Dinosaurs are not the animatronic creatures of *Jurassic Park*, nor are they latex-skinned creatures rooted to the floor, bellowing hideously and repeatedly performing their programmed range of movements in a museum gallery. They are actually known from assortments of fossilized bones: some are known from just a few isolated bones, others from parts of skeletons, and only a few from complete skeletons. "Life" can be given to these collections of bones only through detailed scientific work. Startlingly realistic film trickery used to create the image of a living dinosaur undoubtedly produces a dramatic image, but it also removes some of their mystique. Expectations will continue to rise to the point

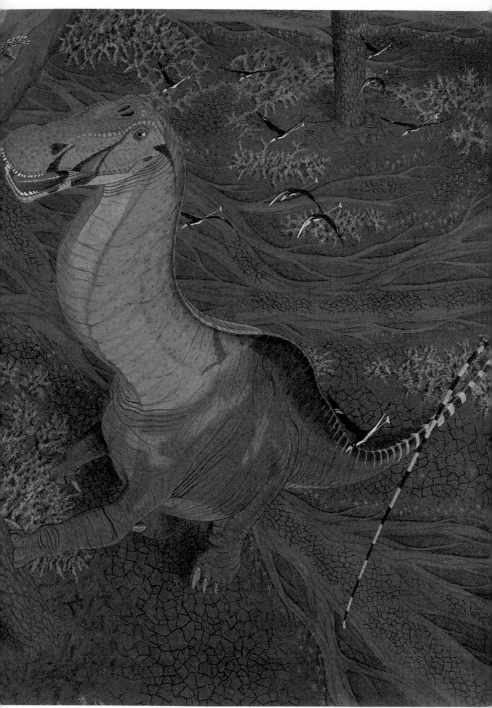

Even with their long necks, apatosaurs may have supplemented their reach by standing on their back legs.

where robotic models in museum galleries will fail to excite children or the public in general – what then? Free-moving robotic dinosaurs kept in "zoos," perhaps, or outdoor dinosaur theme parks. I don't suppose that this can be so far off – dinosaurs have an irresistible commercial edge to them now. I would not be so uncharitable as to critizise such potential progress, but I am concerned that the essential ingredients which have made all this possible – those dry and dusty old fossil bones and the dedicated scientist working in the background – will increasingly be seen as unimportant or irrelevant. I can easily imagine a time when theme park and perhaps even museum administrators feel: "Who cares about the details as long as the public enjoy themselves?"

Well, I care! And I hope you do as well. All the lifelike appearance and movement of dinosaur robots, all the exciting animated sequences of dinosaurs on television and in movies, depends on the ceaseless work of brilliant scientists worldwide, who have brought dinosaurs and their world back to life.

This book was first published in 1991: a very significant year. It was 150 years earlier (in 1841) that Professor Richard Owen addressed the eleventh meeting of the British Association for the Advancement of Science at Plymouth in Devonshire, England. During this meeting, he discussed the types of fossil reptile then known in Britain. While writing his report of the meeting later that year, or perhaps early in 1842, it struck him that three of the fossil reptiles were so completely different from any other known type that they deserved a new name: "Dinosauria." As I will describe a little later, Professor Owen embodied the two characteristics that I have been stressing: scientific brilliance and a fertile imagination. Today as then, scientists still rely on a blend of scientific knowledge and imagination in developing new interpretations or theories. But we have a huge advantage over Professor Owen in all the scientific work that has taken place over the past century and a half, allied to new technologies that allow us to extract information from dinosaur remains, and exchange it with other scientists, in ways unimaginable to the Victorians.

The work of paleontologists – the scientists who work on fossils – is truly a detective story, an unceasing search for new clues to unravel the mysteries of the ancient world. In this book I shall satisfy the curiosity of the reader who has a genuine interest in dinosaurs and wants to know a little bit more. I have attempted to write this book in such a way as to be easily understood by the non-expert. Some subjects are a little difficult to grasp, but I hope that the ideas, as I have explained them, are at least clear. In this I have been helped to an immeasurable degree by scientific colleagues, and the editorial team at Boxtree.

David Norman
Cambridge, England, January 1994

ABOVE: *Far from being slow-moving, pathetic creatures, dinosaurs were vibrantly alive, intelligent, and in some cases quite graceful animals as this illustration of* Ornithomimus *clearly shows.*
RIGHT: *Dinosaur images can be particularly vivid in the minds of children. This painting by a school child from the small village of Esperanza in southern France was prompted by the discovery of dinosaur eggs in the surrounding countryside.*

Introduction

Dinosaur! That name conjures up powerful images in most people's minds. The word was coined only 150 years ago. In that time it has taken on more meanings than its inventor could have dreamed of. It can be an insult: politicians are referred to as "political dinosaurs" to mean that they are stupid, or that their policies are outmoded and they cannot adapt to new ways of thinking. Dinosaurs feature in advertisements as symbols of clumsiness and inefficiency. Never mind that this is nonsense: what is important is that the word "dinosaur" is part of everyday language and creates a powerful image.

Strong as that image is, it is a vague one. Scientists who study the natural world – like all scientists – need to use names which have precise meanings. Without these there would be complete confusion. For example the bird known as a robin in Britain is utterly different from the one which an American knows as a robin. The problem was solved in the mid-eighteenth century by the Swedish naturalist Carl von Linné usually known by the Latin form of his name, Linnaeus. He devised a way of standardizing the names of all creatures so that each one had a unique name, and scientists all over the world would know what was meant by it. The local or common name was kept, but a scientific name was added, chosen from Greek or Latin words. In the eighteenth century Latin and Greek were international languages of science, understood by scholars worldwide.

A simple example of one of these Linnaean scientific names is the one given to humans: *Homo sapiens*. This name comes from the Latin *homo*, "man," and *sapiens*, "wise". It is not the most modest name he could have chosen! But it does reflect the view of eighteenth-century philosophers and scientists. At that time Christian religious views dominated scientific thinking, and man was seen as God's crowning glory of creation.

11

During the early decades of the last century public lectures on subjects such as science, geology and fossils proved immensely popular to an inquisitive audience. In this cartoon from Punch *magazine the young Professor Owen gives a discourse on fossil animals to a packed house.*

So why has the name "dinosaur" become so firmly established in our language and culture? It is, after all, just one of many thousands of names for groups of fossil creatures. One reason is that dinosaurs were the center of one of the first concerted public relations exercises in Victorian Britain, at a time when London was the cultural and scientific center of the world. The other is less easy to define, but I will refer to it as a spiritual factor.

SPREADING THE WORD

The name "dinosaur" comes from two Greek words *deinos*, "terrible", and *sauros*, "lizard." In strict scientific language, the group of animals is called the Dinosauria. That word was first used in the report of a meeting of scientists in Britain, published in 1842. At the time the announcement was probably received with relatively little interest; this was simply the invention of a new name for a poorly known group of large fossil reptiles. The report written by Professor Richard Owen was his contribution to the eleventh meeting of the British Association for the Advancement of Science, an annual forum – still held – at which scientists from a wide range of backgrounds discuss progress in their areas of work. Several years earlier Owen had been given the task of reviewing and reporting on all of the then known

British fossil reptiles. His first report, largely devoted to seagoing reptiles, had been presented in 1839, and this was his second and final report on the mainly land-living creatures. Owen spoke for some two and a half hours and covered a wide range of fossil reptiles, trying to arrange them into some sort of logical or consistent order.

Professor Owen was an extremely ambitious scientist. He worked ceaselessly and very rapidly achieved high status in British scientific circles. His view of dinosaurs was central to his ideas of how life had appeared on Earth, and he was naturally anxious that the idea should be accepted by all. However, in the 1840s he lacked complete fossil skeletons which would allow him to prove beyond doubt what his dinosaurs were really like. In a sense, Owen set the stage for his dinosaurs at the 1841 meeting. Yet he lacked opportunity to show them off to full effect. Then, ten years later, Owen had his chance, and he seized it.

In 1851 the Great Exhibition was opened in Central London's Hyde Park. The centerpiece was an enormous glass pavilion, the Crystal Palace, designed by Sir Joseph Paxton. Inside the gigantic steel and glass construction the products and skills of the British Empire were displayed to the world in all their glory. After a triumphant year the pavilion was dismantled, but it was too well liked to destroy. Instead, it was moved to a permanent site a few miles to the south, in the suburb of Sydenham. During the planning for the reconstruction of the Crystal Palace, which was scheduled for completion in 1854, Albert, the Prince Consort was keen that the grounds around the building should be landscaped and populated with models of prehistoric animals. Prince Albert was well known for his interest in new scientific discoveries. He knew the men who had discovered the early dinosaurs, such as Gideon Mantell and William Buckland, as well as Professor Owen; and he regularly attended the meetings of scientific societies in London. However, it seems that the original idea came from Owen, who had been one of the commissioners for the 1851 exhibition.

The task of reconstructing the prehistoric animals was given to the artist and sculptor Benjamin Waterhouse Hawkins, who had been involved in the construction of the original Crystal Palace, and prior to that had assisted Owen in some of his researches. Owen was appointed Hawkins' scientific adviser. Along with several other prehistoric monsters, four dinosaur models were created and placed in the park. Hawkins and Owen first designed miniature models, which were then used to build full-sized models in Hawkins' workshop. The full-sized models were constructed from bricks, iron columns and hoops, tiles and concrete. Before they were finished a special event was arranged for New Year's Eve 1853. The Crystal Palace Company, with an eye to publicity, arranged for a banquet to be held inside one of the dinosaurs. Hawkins and Owen hosted the gathering, which comprised a mixture of twenty-one scientists and local dignitaries. The organizers did things in style, and the event was widely covered in the newspapers and society journals so that it inspired great curiosity among the general public. The rebuilt Crystal Palace was eventually opened by Queen Victoria in June 1854, and attracted hundreds of thousands of visitors who willingly paid to see not only the glass pavilion and its exhibits, but to gaze in wonder at the prehistoric monsters around it.

There seems little doubt that from this time onward dinosaurs became a part of the English language. So widely did the name and image become accepted that Charles Dickens, the great Victorian novelist (of whom Owen was a particular fan and friend), referred to *Megalosaurus*, one of Owen's dinosaurs, in his 1852 novel *Bleak House*.

The enormous popularity of the dinosaurs in their prehistoric park in London attracted the attention of the Board of Commissioners of Central Park, New York. In 1868 Andrew Green, the far-sighted Board administrator of Central Park, realized the enormous educational value of the Crystal Palace park which, in his words, "...hundreds of thousands of people have visited... annually for the last 15 years." He promptly offered Waterhouse Hawkins a position to design and build a similar prehistoric park in New York's

This contemporary painting shows Paxton's wonderful glass palace as it would have appeared for the re-opening ceremony in the Summer of 1854. Amid the lakes and fountains Owen's prehistoric monsters, including his dinosaurs, must have made a wonderful spectacle.

14

Central Park, to be known as the Paleozoic Museum. This was to show examples of the prehistoric animals that were beginning to be discovered in the United States. Hawkins accepted the job with enthusiasm but, sadly, the Paleozoic Museum was destined never to be built. In 1870 the Central Park Commission was taken over by members of the infamous Tammany Society, whose boss William Marcy Tweed was briefly Commissioner of Public Works before he was convicted of fraud in 1871. Hawkins had already modelled and built many of the prehistoric animals for the museum, but the project was abandoned on the orders of Tweed, and the models and molds smashed and buried. However, Tweed's activities had no long-term effect on the popularity of dinosaurs. A disappointed Hawkins left New York and worked at the College of New Jersey (later to become Princeton University), where he created a number of paintings of prehistoric landscapes based on his work in London and New York. In 1876 he created a new plaster reconstruction of *Hadrosaurus* (now known as *Kritosaurus*) for the centennial Independence celebrations in Philadelphia. Afterwards his model passed to the Smithsonian Institution in Washington, where it was set up as on open-air exhibit. Sadly, this too was not destined to last; *Hadrosaurus* gradually crumbled to dust. But despite the sad fate of Hawkins' dinosaurs, they left their indelible mark in the United States.

Bill of Fare
OF A DINNER GIVEN AT THE
CRYSTAL PALACE,
BY MR. WATERHOUSE HAWKINS
TO PROFESSOR OWEN AND TWENTY GENTLEMEN,

On SATURDAY, DECEMBER 31st, 1853,

IN THE MOULD OF

THE IGUANODON.

Soups.
Mock Turtle. Jullien. Hare.

Fish.
Cod and Oyster Sauce. Fillets of Whiting. Turbot à l'Hollandaise.

Removes.
Roast Turkey. Ham. Raised Pigeon Pie.
Boiled Chicken and Celery Sauce.

Entrées.
Cotelette de Mouton aux Tomâtes. Currie de Lapereau âu Riz.
Salmi de Perdrix. Mayonnaise de Filéts de Sole.

Game.
Pheasants. Woodcocks. Snipes.

Sweets.
Madeira Jelly. Orange Jelly. Bavaroise
Charlotte de Russe. French Pastry. Nougat à la Chantilly.
Buisson de Meringue aux Confiteur.

Dessert.
Grapes. Apples. Pears. Almonds and Raisins. French Plums.
Pines. Filberts. Walnuts, &c. &c.

Wines.
Sherry. Madeira. Port. Moselle. Claret.

CHARLES HIGINBOTHOM, AND EUROPEAN,
ANERLEY TAVERN, MANSION HOUSE STREET.

BANQUET IN *IGUANODON*

TOP LEFT: *The popularization of dinosaurs began in earnest with the rebuilding of the Crystal Palace. A select number of eminent scientists and sponsors of the venture were invited to a banquet in* Iguanodon *on New Year's Eve 1953. The lucky few had invitations written on the wing of a pterodactyl.*

TOP RIGHT: *With seven courses and a variety of wines the banquet must have been an enjoyable affair and no doubt led to many toasts to the two hosts.*

BOTTOM: *It must have been a rather tight squeeze for the 21 guests packed into the mold of the dinosaur. Professor Owen seated inside the head of the dinosaur raises his glass for a toast. The plaques lining the marquee honor the first discoverers of dinosaurs.*

Following the success of the Crystal Palace Park in London, the Commissioners of Central Park, New York, drew up plans for a Paleozoic Museum with Waterhouse Hawkins in charge of the ill fated project. This rare photograph of Hawkins' New York studio shows the models to be well advanced by 1869. Sadly all his hard work was to be for nothing as the project was never completed and the models were destroyed.

THE SPIRITUAL FACTOR?

Dinosaurs have an intrinsic – almost – spiritual – appeal to many people. I suspect that this is because they strike a primeval chord of recognition in us. Many dinosaurs are gigantic, scaly creatures with long tails, sharp claws and teeth – in fact, they resemble the dragon of myth and legend, or the biblical serpent.

Dragons and similar mythical beasts are part of ancient folklore or written tradition in a very broad range of cultures, Chinese traditions of dragons go back over three millennia; ancient Greek myths have their dragons, in America the Toltecs worshipped Quetzalcoatl; there are dragons in Norse myths and European fairy tales. The dragon is almost always a figure of great power, frequently an evil power to be vanquished by the inferior, but resourceful human hero. However, this is not always the case. In China the dragon can be a symbol of energy, and so-called dragon bones and teeth are held in high regard as medicines.

But why does the dragon theme occur at all, why does it recur in so many unrelated cultures, and can it really be linked to our fascination with dinosaurs? I cannot say for sure, but I do have my suspicions. Modern reptiles – particularly crocodiles, lizards, and snakes – are often regarded with horror by adults and children alike, to an extent that is out of all proportion to their true danger.

Dinosaurs are not only reptiles, but are extremely big – staggeringly so in some cases – and all are long dead. They can be awe-inspir-

17

Hawkin's painting of life in the Cretaceous period in New Jersey, inhabited by dinosaurs and mosasaurs, was finished in 1877.

ing when seen in museum displays, but they are not life-threatening; they stand rooted to the spot by the iron framework which supports them. They require one very important ingredient to make them truly terrifying: the human imagination. It is the mind's eye (particularly that of children) which can flesh out the dusty old dinosaur skeleton in the museum and bring it back to life.

For the great majority a childhood fascination with dinosaurs is just a passing craze. However, that interest is too great to die completely, as shown by the frequent appearance of dinosaur-inspired stories in the press or on television. Not a year goes by without some major news story describing a new and fantastic discovery, a new theory to explain their extinction, or some spoof report of a living dinosaur having been spotted in some remote part of the world.

WHY STUDY DINOSAURS?

A paleontologist – and particularly one who studies dinosaurs – is frequently faced with disbelief from others when they learn of his profession; this is shortly followed by a question such as: "Why do you bother to spend your time studying animals that died millions of years ago?"

The answers I can give are many and various. First and foremost, I find them incredibly interesting; and paradoxically, through studying them I learn to appreciate a great deal about the world that I see around me today. This type of work is seen by others as a sort of intellectual escapism – a way of avoiding the huge problems which face our society. While I sympathize with this view, I do not agree with it. There is a small band of dedicated paleontologists scattered around the world, who have devoted their lives to studying and trying to understand all aspects of ancient life – not just dinosaurs. They are historians charting not the history of human civilizations but the far grander history of life on Earth.

In recent years we have become very conscious of our potential to affect the delicate balance of nature – a balance upon which our very existence depends. The world we inhabit today has taken at least 4,500 million years to reach its present state. During this immensity of time the Earth has been altered by geological processes which have heaved and buckled the Earth's surface, and weathering which has gradually worn it down, and by the organisms that have lived before us. Dinosaurs form a part of that history, and quite an important part too: they lasted on Earth as a group for over 150 millions years in one form or another. We have been able to study their rise, flourishing, and eventual decline by collecting their fossilized remains from the rocks. Dinosaurs also help us to learn about the process of evolution, and the complex interactions between these creatures and the Earth that they inhabited. We may be able to learn from the example of groups such as dinosaurs how better to manage this world, and possibly avoid for ourselves the eventual fate of the dinosaurs.

The sheer scale of some dinosaurs can be appreciated in this photograph which shows the paleontologist Jim Jensen lying alongside the shoulder blade of his Supersaurus.

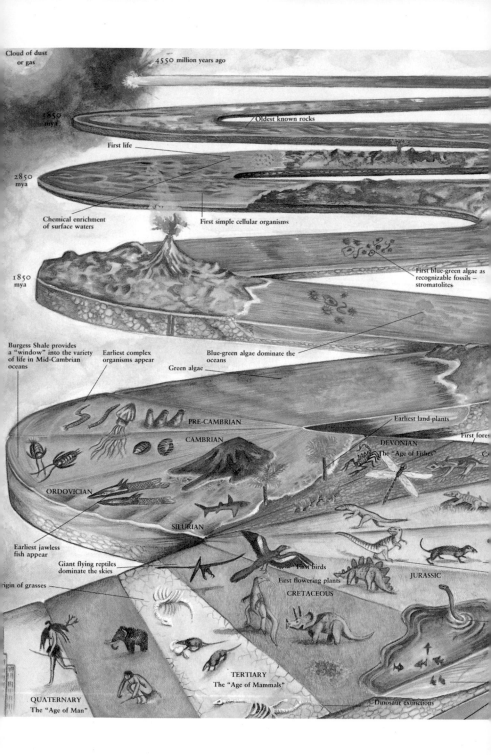

Cloud of dust or gas

4550 million years ago

3850 mya — Oldest known rocks

First life

2850 mya

Chemical enrichment of surface waters

First simple cellular organisms

1850 mya

First blue-green algae as recognizable fossils — stromatolites

Burgess Shale provides a "window" into the variety of life in Mid-Cambrian oceans

Earliest complex organisms appear

Blue-green algae dominate the oceans

Green algae

Earliest land plants

PRE-CAMBRIAN

CAMBRIAN

DEVONIAN
The "Age of Fishes"

First fores

C

ORDOVICIAN

SILURIAN

Earliest jawless fish appear

Giant flying reptiles dominate the skies

First birds

First flowering plants

JURASSIC

rigin of grasses

CRETACEOUS

TERTIARY
The "Age of Mammals"

QUATERNARY
The "Age of Man"

Dinosaur extinctions

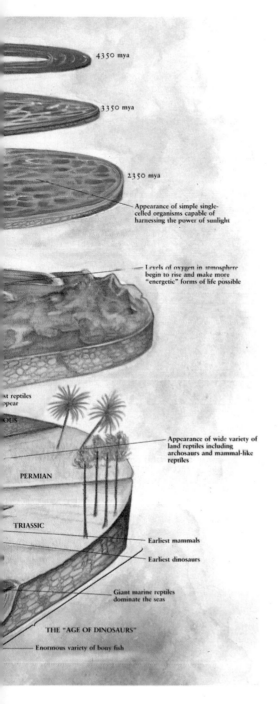

4350 mya

3350 mya

2350 mya

Appearance of simple single-
celled organisms capable of
harnessing the power of sunlight

Levels of oxygen in atmosphere
begin to rise and make more
"energetic" forms of life possible

st reptiles
ppear

OUS

Appearance of wide variety of
land reptiles including
archosaurs and mammal-like
reptiles

PERMIAN

TRIASSIC

Earliest mammals

Earliest dinosaurs

Giant marine reptiles
dominate the seas

THE "AGE OF DINOSAURS"

Enormous variety of bony fish

It is also extremely humbling (and most important for that very reason) to realise that we humans, dominant though we seem at the present time, are destined for extinction – as all groups are in the end. Modern humans have been on earth for little more than 100,000 years, which is a microscopically small period of time in the vastness of Earth's history, and pretty small when compared to the length of time that dinosaurs lasted. We undoubtedly have an important part to play in the story of life on Earth, but it has a future as well as a past. We must therefore understand what has happened in the past, and what we are doing to the planet today, because we are only temporary custodians of the Earth and what we do today may well affect future generations.

Perhaps this is a rather sombre and somewhat introspective tone with which to end this part of the book, but away with the dinosaur "glitz" since they are so often the centre of attraction – for good or bad reasons. Yet for all their wondrous size and variety they are one important group among many which provide us with messages about the history of the Earth. Paleontologists who devote their lives to the study of dinosaurs and a whole host of other fossil forms can tell us the true meaning of those messages.

ABOVE: *Some fossils are extraordinarily beautiful, such as these mineralized ammonites, distant relatives of living squid and octopuses. The coiled shells of these creatures acted as both a casing for the soft body of the animal and a flotation chamber for these sea creatures.*
RIGHT: *One specimen of the dinosaur* Psittacosaurus *has been preserved along with a large cache of pebbles found in the area of its body where its stomach would be expected. This is good evidence for a muscular, food-grinding gizzard.*

CHAPTER ONE

Fossils and Time

Since humans first began to count they have speculated on the age of the Earth. In order to discuss dinosaurs properly, we need some knowledge of the history of the Earth, and of the dinosaurs' place in it. Most people now know that dinosaurs became extinct about 66 million years ago, because it is a subject which has received a considerable airing in the press, on television and the radio. Far fewer know when the first dinosaurs appeared on Earth, or when different types of dinosaurs lived. Fewer still have a clear idea of how old the Earth is. The whole issue of time is further confused today through television and films. The very popular cartoon series *The Flintstones* gives the clear impression that Stone Age man lived alongside dinosaurs, and even had some as pets! It is difficult to erase the mistaken ideas that such scenes create.

The simple fact is that dinosaurs vanished from the face of the Earth almost 66 millions years before modern humans appeared. Our history dates back a mere 100,000 years, so any thoughts of cavemen wrestling with *Tyrannosaurus* are complete nonsense. The ancient Greeks knew that seashells could be found high in mountainous areas, and deduced from this that in the past seas had covered much of the land. Yet while the Greeks and other ancient people generally thought that the Earth had some considerable history, not too many attempts were made to understand or explore that history. Much later, in the sixteenth century, the rise of capitalism and industry spurred on the search for precious metals in the ground to make coinage, and for useful minerals of other kinds. That called for greater attention to rocks, which in turn resulted in the discovery of "figured stones" – what we today would call fossils.

While many regarded these unusually shaped stones simply as freaks of nature and dismissed them as of no importance, others thought they might represent growth of animals within the rocks, into which the breath of life had not been fully blown. Others still, of whom Leonardo da Vinci (1452-1519) is a particularly famous example, were convinced that fossil seashells, which he had collected in the mountains near his home in Italy, were the remains of once living animals which had died on the sea bed and been buried before the land had been raised to its present position.

Fossils

Fossils are one of the keys with which the early philosopher-scientists began to unlock the mysteries of time. To modern scientists they are as important as ever: they underpin all the work that paleontologists do in order to bring back to life ancient creatures such as dinosaurs.

The word "fossil" comes from the Latin *fossilis,* which means "dug up," and in early days the term was used for anything that had been dug from the ground. Nowadays its meaning has become much more restricted and it generally refers to the preserved traces or remains of ancient organisms. In order to understand how fossils are formed, we need to understand several natural processes which occur today, just as they did in the past.

Burial

For a fossil to form, first of all an organism has to become buried. Otherwise, the usual processes of rot and decay, as well as scavenging by other organisms, lead to complete destruction and the return of its bodily chemicals to the environment in which it once lived. That is the natural cycle of nature. Fossil formation actually interrupts or removes from the natural cycle those few organisms destined to become fossils.

Sediment

Burial requires debris – sand, grit, clay, mud. These are provided by weathering. The actions of water and the natural chemicals dissolved in it, and of wind, heat, and cold are capable of wearing down the toughest rocks. Heating and cooling can split rock into fragments, which in turn gradually become worn down into boulders, then pebbles, and eventually through tumbling in streams and rivers are reduced to grains of sand or fine salt. The smallest particles are called sediments. They are carried by rivers down to lakes or to the sea, to be deposited on the bottom in muddy layers. Organisms on a lake or sea floor live in a perpetual rain of fine particles from above which will eventually bury anything that stops moving.

1 Following death, an animal's remains may became fossilized if left undisturbed by scavengers.

2 Skin and flesh rapidly rot as the animal is buried in sediment, leaving just the hard bony parts of the skeleton.

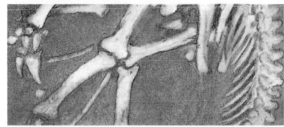

3 Compaction of the sediments around the skeleton locks the skeleton in the ground. Percolating mineral water often alters the chemical composition of the skeleton.

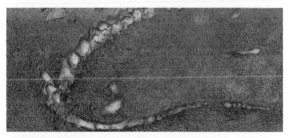

4 Movements of the Earth's crust raise deep rocks to the surface and allow erosion to re-expose the fossilized bones.

For anything to be fossilized a set sequence of events must take place. First, the animal or plant must die. If it was already living in water, or even better on the sea floor, then its chances of being buried are high. Shellfish, since they live in shallow coastal waters, are rapidly buried when they die, and as a consequence they are extremely well represented as fossils. The chances of a land-living creature, especially a dinosaur, becoming buried in this way are much smaller. The carcass would need to be washed (perhaps by a sudden flood when a river burst its banks) downriver into a lake where it could sink in the still water and be buried under sediment. Obviously the longer the time between death and burial, the greater the chances of losing part or all of the carcass through scavenging animals tearing off pieces, or rotted parts falling away. So large creatures such as dinosaurs are very rarely found; when their remains are discovered these are often isolated fragments of bone or single teeth. Very rarely, complete dinosaur skeletons are discovered. Such a find may result after a dinosaur has been buried extremely quickly: trapped in quicksand, caught and drowned in a mudslide or a flash flood, even buried in a sandstorm.

The sequence of events that leads up to the burial of an organism can vary enormously, and through careful excavation and much attention to detail a scientist can learn much about the environment and conditions in which the animal died. This fascinating area of detective work is known as taphonomy (which literally means "the laws of burial"). Great importance is placed not only on the fossil itself, but on the arrangement of its body and how it came to be that way, and the condition of the rocks immediately surrounding it. Many vital taphonomic clues have been lost in past excavations in the general excitement and rush to excavate some new dinosaur skeleton.

Fossilization

Once burial has begun the organism no longer risks being disturbed directly. The "fall" of sediment continues unceasingly and the sediment layer increases in thickness and weight. The softer tissues of the organism continue to rot, leaving only the harder skeletal parts to be preserved, and the sediment becomes packed around the skeleton as the pressure from the overlying sediment continues to rise. Eventually the pressure causes the weaker parts of the skeleton to collapse so that it becomes rather flattened. At the same time the increased pressure presses the particles of sediment ever closer together to form a sedimentary rock. There are various rocks of this kind: some, such as limestones, sandstones, and mudstones are fairly hard; other, such as clays and shales, are soft.

Exposure and discovery

Once sealed in the sedimentary rock the organism is destined to become a fossil. Thus entombed, the fossil will last for ever. But will it be discovered? The last crucial factor depends upon geological processes, and human activity. Geological processes include not only weathering and deposition of sediments, but also uplift, during which the crust of the Earth is bent and buckled upwards. As a result of such upheavals what was formerly a sea floor or lake bottom may be raised to form dry land. If this happens, as it often has in the past, the sedimentary rock will once again become subjected to weathering and erosion, which will finally reveal the fossil.

But as soon as it appears, the fossil begins to disappear. As the rock weathers, so does the fossil, and it too may be rapidly destroyed and lost altogether to science – a belated return to the cycle of nature – unless a paleontologist should come by at the right time to find it.

Fossil types

Fossils can be of a wide variety of types, depending on the conditions which the rock and its entombed fossil have experienced.

Simple preservation The hard skeleton may be preserved practically unaltered for millions of years. Many dinosaur bones retain their chemical makeup, particularly those recently discovered on the north slope of Alaska (see pages 87, 187) which, although they have been discolored to a dark brown, are light in weight and have the texture and chemical composition of normal bone.

Mineralized bone In many cases fossil bones consist not only of the original bone minerals, but also have stone minerals deposited around them in the spaces which would have been occupied by soft tissue within the bone. This is the classic type of heavy, stone-like fossil. It is described as having been petrified, which means "turned to stone".

Natural casts and molds In some cases acidic ground water percolating through the rock in which the fossil is buried may completely dissolve away the original bone. In the space left in the rock new stone minerals may be deposited, producing a perfect stone replica, or natural cast, of the original bone.

Alternatively the hollow space left in the rock may remain unfilled. In this case a paleontogolist may create his own cast of the original by pouring a latex rubber into the natural mold in the rock and pulling out a perfect replica of the original bone.

Dinosaur "mummies" Something that very rarely happens to anything as large as a dinosaur is preservation by burial in windblown sand. The few examples of dinosaurs buried in these very dry conditions are exceptionally interesting because the carcass, instead of rotting as it would normally, dries out and becomes mummified. Impressions of the skin and some of the other soft tissues may be preserved, at least in outline.

Trace fossils These are not the remains of the animals directly preserved, but rather indicate the activity of the animals. They can be of a wide variety of sorts, including evidence of feeding, movement, nesting and eggs, and disease or illness. Their importance cannot be overestimated because, unlike bones, these traces are evidence of what an animal did when it was alive – not how it looked when it died.

FOSSIL CLUES ABOUT DINOSAUR

ABOVE: Footprint traces are evidence of the movement of once-living animals.

ABOVE: Fossil eggs can reveal evidence of nesting and parental behaviour in dinosaurs.

BELOW: The grooves of this Allosaurus *claw show where the nail grew.*

BELOW: A clear impression of the pattern of armor plates embedded in the skin on the back of the ankylosaurian dinosaur, Sauropelta.

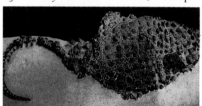

ABOVE: A finely preserved Centrosaurus *skull shows the distinctive nose horn.*

Feeding traces can include bitten leaves, claw or tooth marks left on bones; or at the other end of the process, fossilized dung, which provides direct evidence for the actual diet of the creature.

Movement traces are usually footprints, either individually or in trackways. They are of immense importance to the study of dinosaurs. Such evidence indicates many things about where an animal moves, how fast it could run, whether it moved in social groups or not, and a whole host of other information about the living creature.

Nesting and egg traces are also found, though not often. Eggshell preserves much like bone, though eggs are so fragile that they seldom remain in one piece. However, new discoveries of eggs, nests and even embryo dinosaurs during recent work in the USA, Canada, Europe and Asia have revealed much new information about the social habits, nesting habits, growth and general way of life of dinosaurs.

Disease and injury also leave traces. Pathological bone (bone that shows signs of injury, or is in some other way abnormal) is a very vulnerable indicator of the trials and tribulations of dinosaurs. Calluses where broken bones have mended are reasonably common, and there are some examples of arthritis-like diseases and tumors of various types suggesting dinosaurs enjoyed less than perfect health.

HOW OLD IS THE EARTH?

Early philosophers who accepted that fossils were the remains of once living creatures then found it necessary to explain why, and how, seashells came to be found high on dry land. Many of them saw the fossils as evidence of the Bible story (echoed in Greek and Mesopotamian myth) of the great flood which only Noah and his ark full of animals survived. The Bible had other uses. Read faithfully, the genealogical tables of the Old Testament provided a means of calculating the date of the creation of the world. The most famous calculation of the date is that made by the Irish prelate James Ussher (1581–1656). In 1650 Archbishop Ussher published his conclusion: Sunday October 23, 4004 BC, at 8 o'clock in the morning. Such a conclusion, coming from so eminent a source, commanded great respect at the time. However, to those studying the Earth such a short period of time seemed inconsistent with their own observations. Enormous periods of time seemed necessary to allow for the great thickness of sedimentary rock to form on lake or ocean floors, for mountains to be raised and for rivers to carve out valleys.

The beginnings of a more modern view of a history of the earth came with the Scotsman James Hutton (1726-1797). Hutton has been called "the Father of Geology" because his work formed a starting point for much of the work which is done today. Through studying geological processes – such as weathering, which wears away rocks and even whole mountains, and deposits the debris in the sea; and the uplift of the sea floor to form dry land or even mountains – Hutton realized that the Earth was constantly destroying and renewing itself in what he thought was a never ending cycle. This cycle, he thought, showed "no vestige of a beginning – no prospect of an end."

Hutton's views were published shortly before he died. During his lifetime and into the early decades of the nineteenth century, scientists' view about the age of the Earth came increasingly to reject the strictly biblical interpretations and favour a much longer history, ranging from several tens of thousands of years upwards to many millions of years. They were two main reasons for this. First, careful mapping of rocks and their thickness revealed enormous piles of sediment. In his book, published in 1859, *On the Origin of Species*, Charles Darwin (1809-1882) roughly estimated the thickness of the "geological column" of Britain – that is, the total depth of rock that had formed on the site of the island – to be 13 3/4 miles (22km). An enormous amount of time would have been necessary to build up that amount of sediment. Second, the observation that a bewildering number of creatures had lived and become extinct pointed to the need for a prodigious amount of time to allow this to happen.

In the middle of the nineteenth century the physicist William Thomson (later Lord Kelvin, 1824-1907) attempted to calculate the age of the Earth in a more precise way. Kelvin assumed that the Earth had started as a gas cloud which had condensed to a ball of molten rock, and set out to calculate how long it would have taken to cool to its present temperature. In the 1860s he gave an upper estimate of the age as 400 million years, but progressively whittled this down to just 24 million years by the late 1890s. These dates were still much shorter than the estimates of many geologists of the time, who required an age of many hundreds of millions of years to explain their observations.

Bitter disagreements between physicists and geologists over the age of the Earth were finally overcome by Ernest Rutherford (later Lord Rutherford, 1871-1937), following the discovery of radioactivity in the late 1890s. Radioactivity has been constantly adding heat to the Earth, and therefore making nonsense of Lord Kelvin's estimates which were based on gradual and continuous cooling of the planet in the absence of any other source of heat. Rutherford was able to propose a method of estimating the age of the Earth with far greater accuracy; this involved measuring the rate at which radioactive elements decayed. Today this technique is known as absolute dating.

Absolute dating of rocks can be done by measuring the proportions of radioactive and non-radioactive isotopes of certain elements (an isotope is a form of an element having a particular weight). These isotopes are known to decay (or break down) at an established rate measured in terms of their half-life (the time for half of the radioactive element to decay into a more stable element). Therefore, with a knowledge of the proportions of the unstable radioactive isotope to its stable form, and the rate at which decay occurs, it is possible to date rocks relatively accurately. The most commonly used isotopes are rubidium-87/strontium-87 (where the rubidium decays to form strontium) which is most often used for geological dating; and potassium-40/argon-40 (potassium decaying to argon).

Measured by their strontium ratios, some of the oldest rocks identified on Earth are from Greenland, with an age of very roughly 3,700 million years. These may have formed the first solid surface on the originally molten Earth, about 780 million years after the planet itself formed.

DINOSAURS IN TIME

While the age of the Earth was being estimated during the nineteenth century, much other work was going on. The industrial revolution brought with it a need to discover sources of coal and iron to fuel and build the factories, as well as building materials such as stone, clay, and sand; while canals, railways, and water supplies were being constructed. All of this called for detailed geological knowledge and promoted government-sponsored survey work and the preparation of geological maps over much of the industrial world. Engineers and surveyors such as William Smith (1769-1839) did work of immense value. Smith produced carefully drawn maps showing the exact location of types of rock, and noticed that he was able to compare rocks from one area with another by using their distinctive fossils. Fossils therefore began to be seen as increasingly important as markers which could be used to date, or at least compare, the ages of rocks. Not only the types of fossils in themselves were noted; even more significant were the sequences of types of fossil in different layers of rock of different ages.

Fossil sequences from different areas were studied intently and compared. In some instances similarities were found within overlapping sequences, allowing geologists to join up the rock columns in different areas as a series of successively older or younger rocks. This technique, known as comparative dating, has been indispensable in ordering rocks in local areas. By using this technique William Smith was able in 1815 to publish the first geological map of England and Wales. At about the same time Georges Cuvier (1769-1832) and Alexandre Brongniart (1770-1847) were producing geological maps of the area around Paris, and William Maclure (1767-1840) was mapping the eastern USA.

Comparison of rock types in different parts of the world gradually led to the creation of an internationally accepted geological timescale, which divides the history of the Earth into a series of distinct "bands" or periods: the most recent represented by rocks at the top of an undisturbed sequence, the most ancient by rocks at the bottom. The vastness of Earth history was thus broken up into manageable chunks of time. For the purposes of this book we can skip through the history

Some of the oldest rocks known on Earth are found in Greenland. These date from the Archaean and are more than 3700 million years old. They were probably the first solid surface to form when the Earth's crust began to cool.

of the Earth in order to concentrate on the essential parts that concern the dinosaurs and their times. I shall do this by quickly breaking the history of the Earth into three groups of three time zones: eons, eras, and periods.

Eons

Eons are the largest slices of time: the earliest being the Archaean Eon, a time from which no trace of life is found; the next, called the Proterozoic Eon, when only simple forms of life are found (mainly single-celled organisms); and one which takes us to the present day, the Phanerozoic Eon, which marks the appearance of larger and more complex forms of life. The Archaean ended and the Proterozoic began approximately 2,500 mya (million years ago) – although some very simple fossils have recently been found in Australia which date back some 3,400 mya. Complex animals begin to appear in the fossil records about 600 mya, so this marks the beginning of what for us is the most interesting eon: the Phanerozoic.

Eras

The Phanerozoic Eon is also divided into three blocks of time, called eras. These are the Paleozoic Era (meaning "ancient life," 600-248 mya), a time during which most familiar groups of animals and plants first appeared – with a few notable exceptions such as flowering plants, mammals, and birds.

The Mesozoic Era, often called the "age of Reptiles" (and meaning "middle life," 248-66 mya), saw the arrival of these groups as well as others, and was also marked by the appearance of some imposing groups which are not with us today, including the giant seagoing reptiles such as ichthyosaurs, plesiosaurs and mosasaurs; the giant flying reptiles known as pterosaurs; and, most importantly for us, the giant land-living reptiles, the dinosaurs.

Finally there is the Cenozoic Era, also known as the "Age of Mammals" (and meaning "recent life," 66 mya up to the present), which saw the switch from dominance by reptiles on land, in the air, and in water, to birds in the air, fish and mammals (the whale family and seals) in the sea, and mammals on the land. Of the three eras it is the Mesozoic which concerns us in this book.

Periods

The Mesozoic Era – the time of the dinosaurs – is further subdivided into three periods. The earliest is the Triassic Period (248-208 mya), which started with no dinosaurs (though there were earlier types of reptile), and was dominated by a wide variety of mammal-like reptiles; but by the end of the period some of the earliest dinosaurs appeared

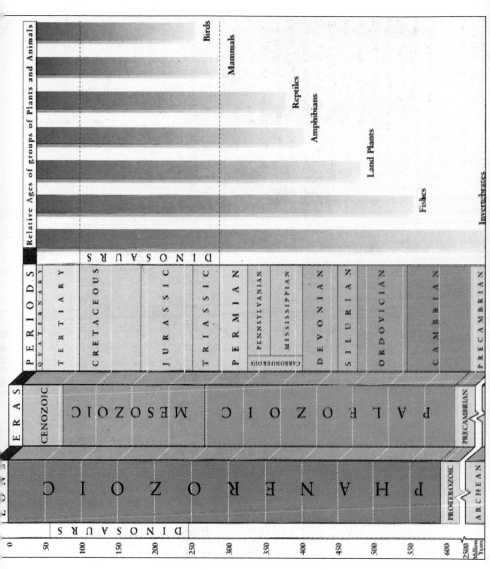

The geological timescale represents our way of dividing up the history of the Earth into manageable portions of time.

– most of which were small, carnivorous types. The Triassic is followed by the Jurassic Period (208-144 mya), during which dinosaurs became very abundant on land, and in which there existed some of the largest creatures that ever walked the earth, the brachiosaurs. Finally there is the Cretaceous Period (144-66 mya), during which dinosaurs became increasingly varied until their sudden extinction at the end of the Cretaceous 66 million years ago.

THE MOVING CONTINENTS

Another essential key to understanding and appreciating the world of the dinosaurs comes from a surprisingly recent revolution in geology: the realization that the continents of the Earth are not fixed in position, but slide slowly and continuously across the surface of the globe. This theory, known as plate tectonics, became generally accepted only in the early 1960s.

Ever since the first reasonably accurate maps of the world were drawn, it must have been obvious that some of the continents would fit snugly together like the pieces of some gigantic jigsaw puzzle. For example the coastline of South America seems to interlock with that of Africa. However, the idea that something as apparently firm and fixed as a continent could move seemed absurd.

In 1912 Alfred Wegener (1880-1930) proposed that the continents had moved during the Earth's history. Back in the Paleozoic, he suggested, the continents were united in a single landmass which he called Pangaea. During the Mesozoic they began to separate, then drift gradually apart. Wegener was able to support this theory with many observations, such as the continuity between separate continents of mountain ranges, geological deposits, and the distributions of fossils. His views were not accepted at the time because nobody could suggest how the continents could move until the late 1950s and early 1960s. At this time new survey work on the sea floor showed that there were areas, known as mid-oceanic ridges in the floor of the ocean where new sea floor was being formed; and other areas known as trenches, where it was being destroyed. The pattern of these ridges and trenches revealed a network of enormous "plates" – known as tectonic plates – which behaved like gigantic, incredibly slow-moving conveyor belts that carried the continents about the surface of the earth. The machines that drive the conveyor belt-like tectonic plates are convection currents in the mantle, the layer beneath the Earth's crust, set up by the intense heat coming from the Earth's core. Despite all the scepticism which greeted Wegener's theory at first, he has been proved correct.

THE MESOZOIC MAP

Just before the Mesozoic, at the very end of the Paleozoic Era, the continents of the world were still united in a single supercontinent, Pangaea.

At the start of the Mesozoic Era (during the Triassic Period) before the first dinosaurs had appeared, Pangaea was still more or less in one piece, so that animals were able to spread widely across the

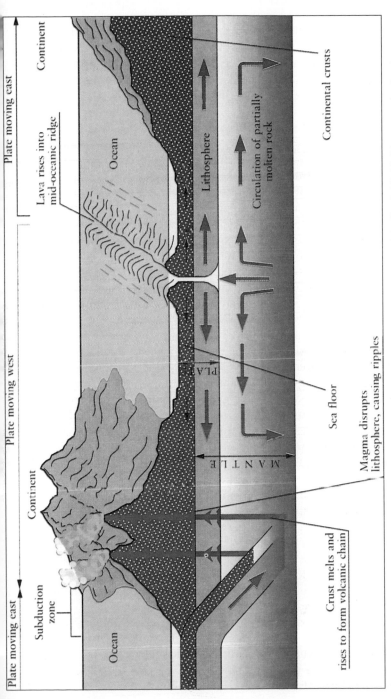

PLATE TECTONICS *Convection currents driven by heat at the center of the Earth drag the crust of the Earth (divided into tectonic plates) apart along spreading 'ridges'. Where plates are forced together, one slides beneath the other forming a "trench". As the crust sinks beneath the surface and re-melts it causes earth tremors and volcanic eruptions at the surface.*

Labels in figure:

Plate moving east

Plate moving west

Plate moving east

Plate moving east

Continent

Lava rises into mid-oceanic ridge

Ocean

Continent

Continent

Ocean

Subduction zone

Crust melts and rises to form volcanic chain

Lithosphere

Circulation of partially molten rock

Continental crusts

Sea floor

Magma disrupts lithosphere, causing ripples

LITHOSPHERE

MANTLE

35

world, unhindered by seas. Indeed the fossil remains of some early Triassic creatures, such as the pig-like *Lystrosaurus*, are found in Australia, Antarctica, South Africa, India and China. At the end of the Triassic, when the first dinosaurs emerge, they too seem to have been widely spread and remarkably alike. For example, small, lightly built carnivorous dinosaurs which look very similar to one another are found in late Triassic rocks in North America, Europe, and southern Africa.

During Triassic times the supercontinents began to break up more completely, and seaways opened to separate the southern continents (South America, Africa, India, Australia, and Antarctica), known collectively as Gondwana, from the northern ones (North America, Europe, and Asia), known as Laurasia. Since the division of the continents was still incomplete, many of the dinosaurs on the separate continents remained very similar. For example, the giant brachiosaurs of North America and Africa are practically identical, as are some of the smaller and more mobile plant eaters. Yet there are already some subtle and curious differences. For example, living alongside the brachiosaurs in southern Africa there is a very spiky stegosaur known as *Kentrosaurus*, which is quite different from the classically large-plated *Stegosaurus* found at the same time in North America. Chinese Dinosaurs also began to take on a rather different appearance from those of the rest of the world; this seems to reflect an early isolation from events elsewhere.

In the Cretaceous continental separation continued apace. Gondwana broke up, with Africa moving away from South America, while North America split away from Europe. India, which had separated from Antarctica at the start of the process, had now reached the Equator. In addition to the general splitting up of the continents, sea levels began to rise. This resulted in the flooding of low-lying areas and the development of a number of shallow seas across several continents. In particular the western and eastern halves of North America were separated by water, as was Europe from Asia; Europe itself was divided into northern and southern parts, and Africa was separated into at least two major land areas.

By the close of the Cretaceous Period the world of the dinosaurs had become highly provincial, with many small, isolated communities separated by seas. This led to an increasing diversity among them as they evolved in isolation.

In this way it is possible to see how the long-term evolutionary history of a large group of animals can be profoundly influenced by factors such as continental movement, even though this might seem to us an incredibly slow process, taking place at a rate of 3/8 to 1 1/5 inches (1 to 4cm) a year.

36

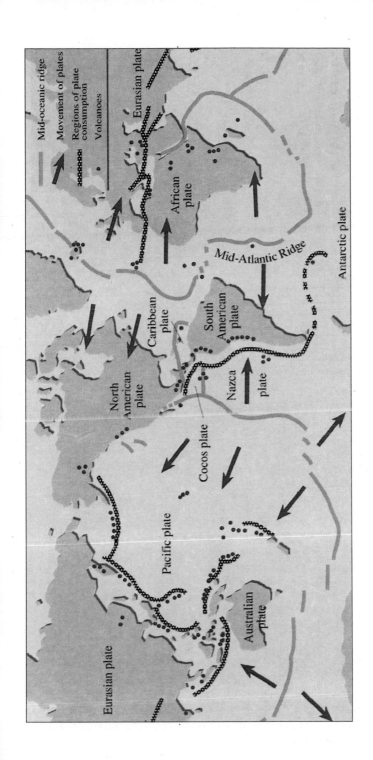

Mid-oceanic ridge
Movement of plates
Regions of plate consumption
Volcanoes

Eurasian plate

African plate

Mid-Atlantic Ridge

Antarctic plate

Caribbean plate

South American plate

North American plate

Nazca plate

Cocos plate

Pacific plate

Australian plate

Eurasian plate

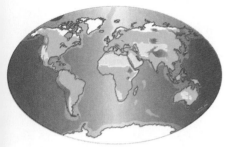

THE CHANGING CONTINENTS

ABOVE: *The present arrangement of the continents is a very stable and familiar pattern to us all, but during the time of the dinosaurs the continents looked very different.*

TOP RIGHT: *During the Triassic period 220 million years ago (mya) all the continents of the Earth were joined together to form a "supercontinent" called Pangaea. With the seas excluded from vast areas of land the climate of the inland areas would have been very dry.*

MIDDLE RIGHT: *During the Jurassic Period, approximately 160 mya, the continents were beginning to separate. Seaways can be seen spreading between continental areas as the separation begins.*

RIGHT: *By Cretaceous times 100 mya, the continents began to become more recognizable as the Atlantic Ocean developed, seaways were generally more extensive, and were in fact about to subdivide many of the continents.*

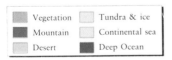

	Vegetation		Tundra & ice
	Mountain		Continental sea
	Desert		Deep Ocean

Triassic Period
The fossil remains of these Triassic creatures were found scattered widely across Pangaea.

Plateosaurus

Ceolophysis

Lystrosaurus
(a dicynodont)

Jurassic Period
As continents separated the dinosaurs diverged in appearance, even though plated dinosaurs were found in both Africa and North America.

Brachiosaurus
(North America/
Africa)

Stegosaurus
(North America)

Kentrosaurus
(Africa)

Yangchuanosaurus (China)

Cretaceous Period
As continents became more isolated, the dinosaurs of certain areas became more distinctive.

Titanosaurus
(Africa/India)

Carnotaurus
(South America)

Triceratops
(North America)

Saichania
(Asia)

PLANTS AND CLIMATE

The climatic conditions of the Mesozoic were much warmer and wetter than those of today. There were no icy polar caps and the weather was surprisingly uniform, with very little difference between summer and winter; there were also incredibly wide latitudinal bands for much of the time, which meant that it was almost as warm in the Arctic as at the Equator.

Triassic weather and vegetation

During the early Triassic Period the temperatures at the poles do not seem to have dipped beneath 10 to 15°C (50 to 60°F) and were somewhat higher at the Equator. However, toward the end of the Triassic the climate seems to have generally become not only hotter, but considerably drier, with a notable spread of arid and desert environments across much of Pangaea. This predominantly hot, arid world is the one that welcomed the very first dinosaurs.

Despite the fact that by the late Triassic the world consisted of a single supercontinent, the plants did not reflect this at the time. In the northern half of the landmass the vegetation was dominated by ginkgo and tree-fern forests, with the forest floor covered by lush fern growth. Farther south, nearer the Equator and in drier conditions, these forests gave way to thinner patchier forests, largely of conifers and cycads.

On the southern continents which were to become Gondwana, many similar trees existed, but they were outnumbered by large seed ferns – such as *Dicroidium*, which appears to have grown to a large deciduous tree. Such plants formed the forest canopy, and smaller fern-like types – such as *Lepidopteris* – made up the ground cover near watercourses.

KEY

1 *Mixed forest of conifer, gingkos and ferns.*
2 *Conifers*
3 *Ferns (Lepidopteris)*
4 *Horsetails*
5 *Tree fern*
6 *Ferns*
7 *Cycad grove*

Triassic scenery

*The world that dinosaurs inherited in the late Triassic was an unfamiliar one.
With all the continents joined together, the inner continental regions away from
the coastal areas would have been hot and dominated by dry, barren areas. This is
borne out by large areas of dry sandy deposits in various parts of the world dating
from this time.*

 *This dry, arid world is seen in the distance, but while much of the land would
have been dry there would have been pockets of rich, highly fertile areas, nearer the
coasts or adjacent to rivers. In these areas a dense cover of vegetation would have
built up, with ferns and horsetails near the ground, and groves of larger conifers,
gingkos, reed-ferns and cycads.*

Jurassic weather and vegetation

During Jurassic times the general environment seems to have become less extreme. The global temperatures fell below those at the end of the Triassic Period, and there was more rainfall, giving rise to more lush and uniform tropical conditions.

Dinosaurs prospered and considerably expanded their range and variety. The favourable climate also led to the appearance of some of the largest land-living creatures that have ever walked on Earth: the gigantic sauropod dinosaurs.

Similar types of plant to those of the Triassic are found in early Jurassic times, but the seed ferns which had dominated Gondwana gradually disappeared and were replaced by cycads and conifers, with ferns covering the ground. The plants found on the southern half of the northern continent seem to have spread at the expense of all others. The existence of Jurassic coalfields, formed from the abundant trees, provides good evidence of the size and thickness of the forests of this time.

Jurassic scenery

The Jurassic world was one which was marked by universal warm conditions, but much greater humidity. Sea levels were rising and the spread of deserts seen in Triassic times was halted and a return to a greener and more lush world is seen. These conditions were ideal for the spread of plants of all types and the establishment of huge forests.

KEY

1 Tree ferns	4 Horsetails	7 Benettitalean cycad
2 Conifers	5 Small tree fern	8 Normal conifers
3 Ferns	6 Ferns	9 Very tall conifers

Cretaceous scenery

The richness of the Jurassic forests did much to encourage browsing dinosaurs. By early Cretaceous times the effect of huge herds of dinosaurs devouring the forests began to show. Into the cleared areas spread the first flowering plants, which were able to grow and reproduce quickly. A recognizably modern forest is beginning to emerge - encouraged by the feeding habits of dinosaurs!

KEY

1 *Mixed conifer forest*	4 *Moss*	7 *Swamp cypresses and Araucaria*
2 *Flowering plants*	5 *Horsetails*	8 *Under storey of broadleaf trees*
3 *Ferns*	6 *Cycad fronds*	

Cretaceous weather and vegetation

During the Cretaceous Period the balmy conditions of the Jurassic continued for a while, but toward the middle of the Cretaceous (about 100 mya) a gradual but accelerating worldwide cooling took place.

For the dinosaurs the Cretaceous seems to have been a very favorable period of time, for although they changed in type – the super-large plant eaters being replaced, for the most part, by smaller, more agile types of dinosaur – they also increased in range and variety. The sudden extinction of all dinosaurs at the end of this period coincides with sudden climatic cooling; but it has not so far been proved possible to confirm that the extinction of the dinosaurs was linked directly to climatic change (see Chapter Seven).

Toward the end of the Jurassic and into the Early Cretaceous a progressive drying appears to have occurred. In general the more equatorial regions appear to have been less heavily forested, and may have developed an almost treeless "savannah" environment, with the ground covered by ferns and horsetails. In the higher latitudes the forests were dominated by conifers, cycads, and ginkgos.

In the Late Cretaceous the first flowering plants appear (the technical name for this group is "angiosperms"). The early forms of these plants were very probably shrubby weeds. In time they increased in range and variety, filling in as smaller trees beneath the forest canopy of conifers. Finally, toward the close of the Cretaceous, they actually formed the canopy as large, broad-leafed trees.

There were no grasses at all during the time of dinosaurs. The dinosaur equivalent of grass was the group cover of ferns.

ABOVE: Tyrannosaurus *such as the two shown here are large, scaly-skinned creatures running on upright legs across a Late Cretaceous landscape. They show all the essential features that paleontologists use to identify dinosaurs.*
RIGHT: *The front foot of a* Brachiosaurus - *an animal that must have weighed close to 30 tons when alive - a marvel of engineering. The toes are short and the bones of the palm form a tubular column ideal for supporting immense weight.*

CHAPTER TWO

What are Dinosaurs?

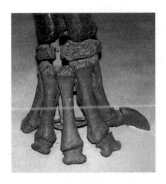

Nearly everyone knows what some dinosaurs look like, such as *Tyrannosaurus, Triceratops* and *Stegosaurus.* But they may be much more vague about the lesser known ones, and may have difficulty in distinguishing between dinosaurs and other types of prehistoric creatures. It is not at all unusual to overhear an adult, taking a group of children around a museum display, being reprimanded sharply by the youngsters for failing to realize that a woolly mammoth was not a dinosaur, or – more forgivably – that a giant flying reptile such as *Pteranodon*, which lived at the time of the dinosaurs, was not a dinosaur either.

So what exactly is a dinosaur? And how do paleontologists decide on the groups they belong to? There are four features that determine what counts as a dinosaur:

Dinosaurs lived only in the Mesozoic Era. Dinosaur remains have been discovered in rocks which range in age from the latest part of the Triassic Period (about 220 mya) throughout the Jurassic Period and up to the end of the Cretaceous Period (66 mya) – spanning approximately 155 million years of Earth history. Therefore any prehistoric creature which comes from rocks which can be dated at older then 220 million years or younger than 66 million years is very unlikely to be a dinosaur. This immediately excludes an enormous number of types of fossil, and of course rules out the woolly mammoth, whose remains may be a mere 100,000 years old.

Dinosaurs are reptiles. Living reptiles are recognized most easily by their scaly skin, and the fact that they lay eggs with shells. These are rather difficult characteristics to identify in practice, because only a few dinosaurs have been preserved with skin impressions or with evidence of egg shells. Nevertheless it does help us to eliminate quite a wide range of organisms with which they may conceivably become confused. For example it excludes fishes, amphibians (which lay eggs without shells) birds (with feathers), and mammals (with hairs, and live-born young).

All dinosaurs lived only on land. This is a particularly useful characteristic because it helps to eliminate a wide range of giant and often bizarre fossil reptiles which are most frequently confused with dinosaurs. Among these are the giant seagoing reptiles of the Mesozoic Era, the plesiosaurs, ichthyosaurs, and mosasaurs; as well as the varied flying reptiles, the pterosaurs, which ranged in size from the size of sparrows, like *Pterodactylus*, to that of small aircraft, such as *Quetzalcoatlus*.

All dinosaurs walk on upright, pillar-like legs. That is the final and most distinctive characteristic of all. Among the reptiles, only dinosaurs have managed to arrange their legs in such a way that they are held directly beneath their body. All other reptiles have their legs splayed outward at an angle from the sides of the body, so that their feet are wide apart. This particular arrangement of dinosaurs' legs is very similar to that seen in living birds and mammals – a feature that was first noticed by Richard Owen in 1842. For the dinosaurs it was doubly advantageous: it allowed the legs to swing most efficiently

DINOSAUR LIMBS: *"Sprawlers" (left), such as amphibians and the majority of reptiles, walk with their legs splayed and the belly close to the ground.*
"Pillars" include dinosaurs, such as this apatosaur. Their legs are tucked beneath their bodies to best support their great weight. This arrangement is very similar to that seen in birds and mammals.

beneath the body, so that they could have long-striding legs and could run fast; and it also meant that the legs were able to support the great body weight of the largest dinosaurs, because they could act literally like supporting pillars.

Armed with this list of four identifying characteristics, you should never be caught out again. But there is just one final warning. Dinosaurs may be big, but don't depend on that! I have had the privilege of holding a complete dinosaur skeleton in the palm of my hand – the tiny *Mussaurus* (literally "mouse lizard") from Patagonia. This may be an extreme example, and the creature was quite probably a young baby; but nevertheless there were a number of small dinosaurs – roughly as large as a medium-sized dog – ready to catch out the unwary museum visitor, or trip up the potential dinosaur expert.

DINOSAUR NAMES

Just as there is a range and variety of animals living today, which we recognize as belonging to different groups, such as horses, cats, dogs, pigs, sheep, or squirrels, so it is with dinosaurs. But dinosaurs do not have common names with which everyone is familiar, such as "horse," "cat," or "dog," to distinguish them. Instead we have used their scientific names.

To explain scientific names let me use the examples I have already given of common animals. The horse has the scientific name *Equus caballus*, the dog is *Canis familiaris*, and the cat *Felis Catus*. Each scientific name consists of two parts: the first name (always written with a capital letter) is the generic name for the animal; for example *Felis* is the name of a genus (group) of cat-like carnivores which also includes the wildcat, *Felis silvestris*. The second name (always written with a small letter) is the species name, and the one that is special to that type of animal. Only the domestic cat belongs to the species *catus*. This is an internationally recognized system for identifying and naming species. The names are often repetitive: for example both *Felis* and *catus* simply mean "cat".

Dinosaurs are named in the same way. To take a well-known example, *Tyrannosaurus rex* literally means "king-lizard king." Often species names are not only repetitive but unnecessary. Among the dinosaurs, a genus may well comprise only one species – there may have been others which have not been found. I will almost always refer to a dinosaur by its generic name alone, so *Tyrannosaurus rex* will be referred to as *Tyrannosaurus*. If a species name is used, this will be to avoid confusion, or because there is something special about that name.

49

CLASSIFYING DINOSAURS

Since 1842, when Richard Owen took the vital first step of recognizing dinosaurs as a distinct group, several hundred species have been listed, varying greatly in size and shape. Having to remember each one in great detail would be very difficult. Fortunately they are not all completely different, but fall into related groups of families which can be arranged more understandably. Much of the classification is done by studying the creatures' legs, hip bones, and feet. Since dinosaurs, unlike any modern reptiles, walk with their legs tucked underneath their bodies, these bones are highly distinctive – as will be discussed in detail in Chapter Five.

A great step in understanding dinosaur relationships came in 1887. By this time a number of reasonable skeletons of dinosaurs had been described, and Professor Harry Govier Seely of King's College, London, noticed that they could be separated into two distinct groups on the basis of differences in the structure of their hip bones. Seely called these two groups the Saurischia (meaning "reptile-hipped") and the Ornithischia (meaning "bird-hipped").

The Saurischians

As the name suggests, the hip bones are arranged in much the same way as they are in other reptiles. The large, blade-like upper bone, called the ilium, is connected to the backbone by a row of strong ribs, and its lower edge forms the upper part of the hip socket. Beneath the ilium there is a large bone which points downward and slightly forward – the pubis – and behind this there is a bone extending backwards – the ischium. All three bones meet at the hip socket, which forms a deep, round opening in the side of the pelvis. Large, powerful leg muscles are attached to each of these bones. The dinosaurs which have this type of hip structure fall into two distinct types.

Theropods include all the carnivorous (meat-eating) types; the name means "beast foot" on account of the very sharply clawed three-toed feet of these animals. The group includes such notable dinosaurs as the giant *Tyrannosaurus*; the small and very agile *Deinonychus*; the very early form *Coelophysis*; the mysterious new dinosaur recently excavated in Britain, *Baryonyx*, with its massive claws; and even some toothless types such as *Oviraptor* and *Struthiomimus*.

The body form of all theropods tends to be very similar: long, powerful hind limbs ending in sharply clawed bird-like feet; slender or lightly built arms; a chest which is rather short and compact; a body balanced at the hip by a long, muscular tail; a neck which tends to be sharply curved and very flexible; and a head equipped with large eyes

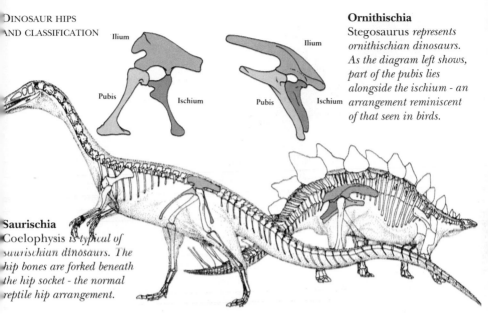

Ornithischia
Stegosaurus *represents*
ornithischian dinosaurs.
As the diagram left shows,
part of the pubis lies
alongside the ischium - an
arrangement reminiscent
of that seen in birds.

Ilium

Ilium

Pubis

Ischium

Pubis

Ischium

Saurischia
Coelophysis is typical of
saurischian dinosaurs. The
hip bones are forked beneath
the hip socket - the normal
reptile hip arrangement.

and long jaws, nearly always lined with dagger-like teeth. Despite this overall similarity there are a number of distinct types.

Some of the best known are the carnosaurs, the classically huge predatory animals including *Allosaurus, Megalosaurus, Carnotaurus, Tyrannosaurus, Tarbosaurus, Albertosaurus,* and an apparent dwarf tyrannosaur which has been named *Nannotyrannus*. All of these types typically have huge, powerful heads mounted upon very thick, powerful necks, and they often have rather small arms for their size.

Another group is a little more varied, and rather rarer in the fossil record. Known as ceratosaurs, they include the very early theropods *Coelophysis* and *Syntarsus*, and *Dilophosaurus* with puzzling thin, bony crests on its head. *Ceratosaurus* is also adorned, but this time with a horn on its nose.

Another, rather mixed bag of theropods I will call coelurosaurs; all of these are generally rather slender, lightly built creatures with long, flexible necks and small heads, and rather long arms and hands. They include the tiny *Compsognathus*, the toothless (and probably omnivorous) *Ornithomimus*, the toothless and extremely odd-looking *Oviraptor*, and more distantly related creatures such as *Troödon, Dromaeosaurus, Deinonychus* and *Ornitholestes*. Though by no means a natural group, they show the lightness and graceful stance of some theropods, and demonstrate more clearly than most a theory which I shall discuss later: that the birds may have evolved from such creatures.

Finally there is a range of theropods most of which have been discovered in recent years, and which defy clear analysis. They are prob-

Ceratosaurs
Dilophosaurus *is a large representative of this group of carnivores, several of which had horns or crests.*

Carnosaurs
Daspletosaurus *is notable for its huge powerful head which enabled it to feed on many of the large herbivores.*

ably best described as "odd theropods." Examples include *Baryonyx*; a gigantic pair of arms belonging to a creature known as *Deinocheirus*; and a possibly bird-like theropod, *Avimimus*. Lastly there is a group of theropod fossils from Mongolia which have been given various names and may represent a totally puzzling group known as segnosaurs. It is not even certain if these are theropods, but whatever else they are they are strange. They have partly toothless beaks, enormous sickle-shaped claws 2ft (60cm) long or more on their hands, broad feet, and a pelvis which looks very similar to that of an ornithischian dinosaur.

Although the body form of these animals is very similar in nearly all cases, the theropods were extraordinarily successful as a group. Proof of their success is demonstrated by the likelihood that their descendants may still be with us today. It seems highly likely that modern birds are derived from one of these groups of theropod dinosaurs.

Ornitholestes, *a larger and more aggressive carnivore, is still light, well-balanced and highly maneuverable.*

"Coelurosaurs" Compsognathus, *one of the smallest members of this group of dinosaurs, is known to have caught lizards.*

"Odd theropods" Baryonyx, *a recently discovered theropod presents numerous problems. Some believe it may have eaten fish.*

Sauropodomorphs, in contrast, are all herbivores – that is, plant-eaters. They range in size from the diminutive forms, generally known as prosauropods ("early sauropods"), which appear in the Late Triassic and Early Jurassic, up to the gigantic sauropods of the Jurassic and Cretaceous Periods.

The prosauropods include forms such as *Anchisaurus, Massospondylus, Riojasaurus, Mussaurus, Plateosaurus, Lufengosaurus,* and *Efraasia*. Most of these are medium-sized creatures 13 to 20 ft (4 to 6m) long which are capable of walking on all fours or on their hind legs alone; but a few show early signs of becoming larger and much heavier, and seem entirely four-footed like the later sauropods. *Mussaurus* is considerably smaller than the others as it name, which means "mouse lizard," suggests – although only juvenile specimens have been found so far and the adults would have been several times larger.

The sauropods are the true giants of the Mesozoic Era and include such notable creatures as *Diplodocus, Apatosaurus* (better known as *Brontosaurus*), *Dicraeosaurus,* and *Cetiosauriscus*. All of these seem to

RIGHT: **"Prosauropods"**
Lufengosaurus *is a large bodied early sauropodomorph; note that it can still balance on its hind legs.*

BELOW: Saltasaurus *is one of a variety of unusual sauropods. This one has peculiar bony armor studding its back and tail.*

CENTRE: **"Brachiosaurs"**
Brachiosaurus *is typical of the group, with long necks, shorter tails and long front legs giving them a great reach.*

FAR RIGHT: **Diplodocids**
These are the giants, such as Apatosaurus *seen here, with enormously long necks and tails.*

belong to the same related group, and tend to have long, slender bodies, whip-like tails, long, shallow faces, and thin, pencil-shaped teeth.

Those in another group, which includes *Brachiosaurus, Camarasaurus, Euhelopus,* and *Opisthocoelicaudia*, again seem to be related to one another. In contrast, they tend to be rather more compact animals, with high-shouldered bodies, shorter tails, shorter and high snouted heads, and much larger teeth. In addition to these types, there are various unusual sauropods: *Saltasaurus* from Argentina has curious armor plating over its back and sides; *Shunosaurus* from China seems to have had a bony club on the end of its tail; *Mamenchisaurus* and *Barosaurus* seem to

54

have had extraordinarily long necks relative to their bodies; and *Magyarosaurus* seems to have been a rare miniature sauropod.

The heyday of the sauropodomorphs was evidently the Jurassic, and they were both abundant and diverse in most parts of the world. In Cretaceous times the group was still to be found, but its members seem to have made up a small part of any community instead of being the dominant type.

The Ornithischians

The arrangement of the bones in the hips of these dinosaurs is, as the name suggests, very similar to that seen in living birds – though, confusingly, there is no family link with birds. While the ilium and ischium bones are arranged very similarly to the saurischian dinosaur, the pubis, instead of pointing downward and a little forward, is a narrow, rod-shaped bone which lies alongside the ischium. This pattern becomes a little obscure in some ornithischians, especially among those from the later Cretaceous Period (such as the ceratopians and ankylosaurs), through shortening of the pubis and growth of a new forwardly pointing part to the bone; but the pattern is still evident.

In addition to this difference in hip bones, there are other features of this group which are not found in saurischians. All ornithischians seem to possess a small horn-covered beak perched on the tip of the lower jaw (a very distinctive feature of dinosaur illustrations). Somewhat less obviously, they have rows of long bony rods packed

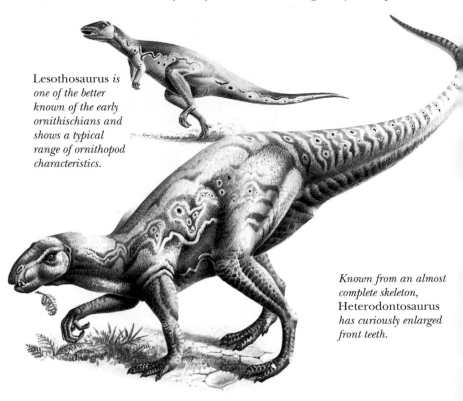

Lesothosaurus *is one of the better known of the early ornithischians and shows a typical range of ornithopod characteristics.*

Known from an almost complete skeleton, Heterodontosaurus *has curiously enlarged front teeth.*

56

along the sides of the spines on the backbone, which helped to stiffen and strengthen the back; these are sometimes visible on museum skeletons.

Ornithischians were, in contrast to saurischians, entirely herbivorous; they are also considerably more varied in appearance then the saurischians. There are five major groups.

Ornithopods include many small to medium-sized animals which ran on their hind legs for most of the time. Examples of these include *Lesothosaurus, Heterodontosaurus, Hypsilophodon, Dryosaurus, Rhabdodon* and *Yandusaurus*, all of which are small forms, none more than about 10 ft (3m) long. This type of dinosaur is found throughout the Mesozoic and seems to have been one of the most successful small herbivore groups. Medium sized types include *Iguanodon, Tenontosaurus, Camptosaurus*, and *Ouranosaurus*; such creatures reached lengths of about 10 meters and were particularly abundant in the early part of the Cretaceous Period. However, in the Late Cretaceous there appeared another group known as hadrosaurs or, more popularly, as duck-billed dinosaurs. These grew to lengths of 43 ft (13m) in some cases. They became very diverse. Some seem to have lived in very large herds, and were evidently highly social creatures. They were also extremely efficient herbivores, with special grinding teeth and muscular cheeks. In some respects the huge herds of hadrosaurs inhabiting the plains of North America in the Late Cretaceous seem to be equivalent to the hordes of buffalo seen in the past on the North American plains, and wildebeest of the African plains.

Ceratopians, the distinctively horned and frilled dinosaurs, with curiously narrow, parrot-like beaks, appeared very late in dinosaur history, only during the second half of the Cretaceous Period. They ranged from the small, rather ornithopod-like *Psittacosaurus, Protoceratops* – whose eggs were first found in Mongolia in the 1920s – *Leptoceratops, Avaceratops* and *Bagaceratops*, to the massive, rhinoceros-like *Centrosaurus, Triceratops, Styracosaurus, Anchiceratops, Chasmosaurus* and *Torosaurus.*

The first ceratopians appear in the middle of the Cretaceous Period in Asia and evolved extremely rapidly to become one of the most abundant and diverse groups of Late Cretaceous dinosaur. Like the hadrosaurs, these dinosaurs also became phenomenally abundant, as we know from the existence of massive ceratopian "graveyards" at some localities. It seems likely that they too lived in large herds which roamed the plains of the Northern Hemisphere. The sharp, hooked beak formed a clean cutting tool for feeding on plants,

Triceratops *was a massive type showing the well-developed horns and frill that characterize the group.*

Protoceratops *is a small ceratopian that was particularly common in Asia during the late Cretaceous.*

and behind this the jaws were lined by dense rows of teeth, forming guillotine-like blades which could have sliced up the toughest plants. The horns and frills which adorn the heads of many of these creatures may have had a number of uses. One obvious purpose of the horns may have been to defend themselves against predatory dinosaurs, but various other uses relating to their behavior have also been suggested, as we shall see later.

Pachycephalosaurs are rather poorly known creatures; they resemble ornithopods in their body proportions but have most distinctive, curiously domed and massively reinforced heads. These forms appear to have arisen during the middle part of the Cretaceous and persisted to the close of that period, but they

Stegosaurs
The huge plated dinosaur
Stegosaurus *is a classic example of this type of dinosaur.*

Pachycephalosaurus
As this Stegoceras *shows, these dinosaurs looked rather similar to ornithopods except for their domed heads.*

58

were never particularly abundant as a group. It has been suggested that they lived in relatively inaccessible places, such as upland areas, where their remains are less likely to have become fossilized.

The creatures had the appearance of being the Mesozoic intellectuals; however, the bulge of the head was not for brains but filled with solid bone! It seems likely that the thickening of the skull bones occurred because they used their heads as battering devices.

Stegosaurs, the well-known plated dinosaurs of which *Stegosaurus* is the famous example, seem to have lived almost exclusively during the Jurassic, with some fragmentary reports from the early part of the Cretaceous Period only. The structure of the spines and plates of these animals has proved to be extremely interesting, and provides

Ankylosaurus
This example is Euoplocephalus, *a heavily built, formidable armored dinosaur from North America.*

Pinancosaurus *from Asia has a large tail club which it swung at the legs of predatory tyrannosaurs.*

59

evidence in the debate as to whether dinosaurs were "warm-blooded" or "cold-blooded" (see pages 171, 253).

Ankylosaurs are the armoured tanks of the dinosaur world. These amazing creatures were fully covered in thick bony plates in order to ward off the attentions of the larger theropods. They appear early in dinosaur history with *Scutellosaurus* and *Scelidosaurus* in the Early Jurassic, but become abundant only in the Late Cretaceous of Asia and North America.

Dinosaur family trees

The major groups of dinosaurs listed above can be grouped together into a family tree in order to show their relationships more clearly. The advantage of this type of display is that it is quite easy to see when certain groups of dinosaurs lived, and how long they lived compared with other groups. It is also possible to see that not all types of dinosaur fit neatly into the groups mentioned above. We are still not entirely clear about the relationships of all dinosaurs, especially some of the newer and puzzling discoveries, and there are a few problems with this family free.

The early dinosaurs give rise to several difficulties. As mentioned earlier, the first recognizable dinosaurs appear in the Late Triassic, between 220–215 mya, and are medium-sized predatory types such as *Herrerasaurus* and *Staurikosaurus*. To date, these two are not well enough known to be able to put them into one of the better recognized groups of dinosaurs, for example with other theropods. Further recent discoveries of *Herrerasaurus* in Argentina may change this situation soon. For the moment it is probably best to class these types of dinosaur under the vague term "protodinosaurs" until things become clearer. Another puzzling, and still rather poorly known creature is *Pisanosaurus*, which may be a very early ornithischian. A jaw with some teeth which are clearly those of a plant eater had been found, and some rather puzzling parts of a very slender skeleton.

Thinking about the idea of "protodinosaurs," it seems to me quite reasonable to suppose that some of the very earliest creatures that we recognize as dinosaurs may indeed have been ones that belonged neither to the Ornithischia nor the Saurischia, and that these latter groups only developed later.

The segnosaurs are an extremely puzzling group of dinosaurs first discovered in Mongolia in 1979. Their many strange features seem to link them variously with ornithischians, theropods, and sauropodomorphs. They have been placed here between the sauropodomorphs and the theropods because that is the majority view among experts at the moment.

The relationship of birds to dinosaurs was for long a controversial subject. Although it is no longer in serious doubt (see page 198), it should be noted that the birds have their closest family links with the carnivorous theropod dinosaurs belonging to the saurischian group, rather than the ornithischian ("bird hipped") dinosaurs as might have been expected. Their nearest relatives appear to be found among the group of which the highly predaceous *Deininychus* was a member – the dromaeosaurids.

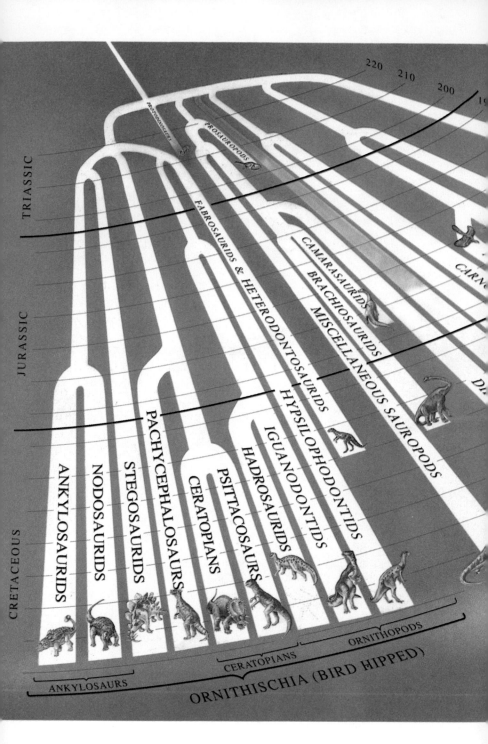

220 210 200

TRIASSIC

JURASSIC

CRETACEOUS

PRODTOMMOSAURS

PROSAUROPODS

FABROSAURIDS & HETERODONTOSAURIDS

CAMARASAURIDS

BRACHIOSAURIDS

MISCELLANEOUS SAUROPODS

CARN

HYPSILOPHODONTIDS

IGUANODONTIDS

HADROSAURIDS

PSITTACOSAURS

CERATOPIANS

PACHYCEPHALOSAURS

STEGOSAURIDS

NODOSAURIDS

ANKYLOSAURIDS

ANKYLOSAURS

CERATOPIANS

ORNITHOPODS

ORNITHISCHIA (BIRD HIPPED)

62

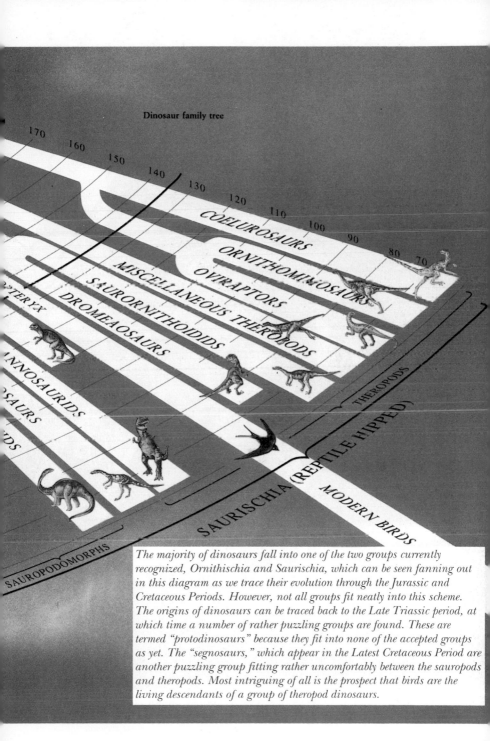

Dinosaur family tree

170 160 150 140 130 120 110 100 90 80 70

COELUROSAURS

ORNITHOMIMOSAURS

OVIRAPTORS

MISCELLANEOUS THEROPODS

SAURORNITHOIDIDS

DROMEAOSAURS

...TERYX

...NNOSAURIDS

...SAURS

...DS

THEROPODS

SAURISCHIA (REPTILE HIPPED)

MODERN BIRDS

SAUROPODOMORPHS

The majority of dinosaurs fall into one of the two groups currently recognized, Ornithischia and Saurischia, which can be seen fanning out in this diagram as we trace their evolution through the Jurassic and Cretaceous Periods. However, not all groups fit neatly into this scheme. The origins of dinosaurs can be traced back to the Late Triassic period, at which time a number of rather puzzling groups are found. These are termed "protodinosaurs" because they fit into none of the accepted groups as yet. The "segnosaurs," which appear in the Latest Cretaceous Period are another puzzling group fitting rather uncomfortably between the sauropods and theropods. Most intriguing of all is the prospect that birds are the living descendants of a group of theropod dinosaurs.

ABOVE: *During the early 1880s the* Iguanodon *skeletons collected at Bernissart were painstakingly reassembled. A wooden scaffold, from which bones were suspended by ropes, was used to mount the first complete skeleton.*

RIGHT: *Robert Plot's illustration is clearly of a dinosaur bone, the broken upper part revealing the marrow cavity of the original fossil. In 1677 however, dinosaurs were unknown and the bone was mistakenly labelled "Scrotum Humanum".*

CHAPTER THREE

Discovery of the Dinosaurs

Long before dinosaurs were seriously studied, people dug up their remains. Dr Dong Zhiming, a noted authority on Chinese dinosaurs who works at the Institute for Vertebrate Paleontology and Paleoanthropology in Beijing, has recently reported that dinosaur teeth (which the Chinese called "dragon's teeth") were known in the sixteenth century BC; and there are even written records of "dragon bones" having been found during the third century AD in areas of China that are now known, as a result of modern research, to be rich in dinosaur remains.

Long before Professor Richard Owen recognized dinosaurs as a group and gave them their name, the road to his discovery was being paved by other people. Although they had little idea of what they were dealing with, we should not ignore their work. They were highly intelligent, inquisitive, and energetic people struggling to make sense of fragmentary fossils.

Guesses about bones

In 1677 Robert Plot, who was Professor of what was then called "Chymistry" at Oxford University, published a book entitled *The History of Oxfordshire, being an essay toward the Natural History of England*. In his book, Plot described the lower end of a gigantic thigh bone (the part which would have formed the upper half of the knee joint) which had been dug out of a quarry at Cornwell in Oxfordshire. Plot described the bone at some length, ruling out any possibility that it might be some sort of stony artefact because it revealed across the upper end the spongy structure typical of bone, as well as an internal marrow cavity. Having satisfied himself that it must be a real bone, but one which had been petrified, he tried to identify the creature to which it belonged. He concluded that "...it must have belong'd to some greater *Animal* than either an *Ox* or *Horse*, and if so (say almost all other *Authors* in the like Case) in probability it must have been the *Bone* of some *Elephant*, brought hither during the Government of the *Romans* in Britain." Even this conclusion he doubted because he could find no historical records of the Romans bringing elephants to Britain.

But he was able to test his hypothesis in true scientific manner, for during 1676 the skeleton of an elephant was brought to Oxford and he was able to examine it. Its bones proved to be different from the specimen found at Cornwell, and he therefore concluded that this bone must have been the thigh bone of a human giant. Plot's illustration of the bone allows us to identify it as the lower end of the thigh bone of the carnivorous dinosaur *Megalosaurus*, whose bones have been found in Middle Jurassic rocks in Oxfordshire.

A French abbot named Dicquemare, who lived near the Normandy coast and collected natural curiosities as a hobby, published a report in 1776 on some fossils which he had collected from the foot of the sea cliffs known locally as the Vaches noires ("Black cows"). Among the various shells and bony fragments he seems to have discovered a dinosaur leg bone, as far as can be judged by the very detailed description which he wrote. Unfortunately, he provided no illustration, and the bone can no longer be traced. In succeeding years further specimens were collected from the Normandy coast by another priest named Bachelet, who donated his collections to the National Museum in Paris. These specimens were described and illustrated in 1808 by Georges Cuvier (of whom more will be said shortly) as the remains of two distinct types of fossil crocodile. One of them later turned out to be the remains of a dinosaur. Professor Philippe Taquet of the National Museum of Paris has rediscovered these specimens among the collections of his museum – a row of vertebrae from the backbone of the Late Jurassic carnivorous dinosaur *Streptospondylus*.

Similar discoveries were also being made on the other side of the Atlantic during the late eighteenth century. Large bones were found in New Jersey, and in 1787 Dr Caspar Wistar and Timothy Matlack reported on them to the American Philosophical Society in Philadelphia. Hadrosaurian dinosaurs have since been discovered in rocks from this area, so it seems quite likely that the finds were dinosaur bones. Similarly in 1806 William Clark, the explorer, recorded in his diary the discovery of what was almost certainly a dinosaur leg bone in a river bank near Billings, Montana – again an area which has since proved to be rich in dinosaur remains. But the first irrefutable remains of dinosaurs from America were not reported and described until 1820. They were found in Early Jurassic rocks in the Connecticut Valley, and were reported to be human remains in the *American Journal of Science* by one Solomon Ellsworth Jr. The bones are still preserved in the collections of the Yale Peabody Museum, and can be identified as belonging to the early dinosaur *Anchisaurus*.

Early footprints

Another equally curious discovery to be found in the historical literature is that of large, three-toed, and distinctly bird-like footprints found by Pliny Moody on his property in the town of South Hadley, Massachusetts. After languishing unappreciated, except as a local curiosity, for more than three decades, these prints became the subject of detailed analysis carried out between the mid-1830s and the mid-1860s by the Reverend Professor Edward Hitchcock of Amherst College. Hitchcock supported the view that the many thousand footprints which he collected and studied belonged to gigantic ancient birds, though it is now apparent that they were made by dinosaurs.

Cuvier's scientific method

The first really thorough study of fossil bones was carried out by the famous French scientist Baron Georges Cuvier (1773-1838). From 1799 onward he worked in Paris as an anatomist at the Jardin des Plantes, otherwise known as the National Museum of Natural History. One of the most brilliant and radical thinkers of the time, he believed that all forms of animal life conformed to a limited number of pat-

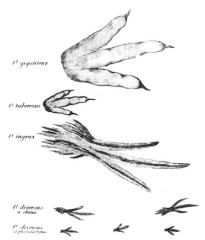

These very distinctive three-toed footprints collected from the Connecticut Valley by Edward Hitchcock were so bird-like that their origins were not seriously doubted for many decades. First found by Pliny Moody in 1802, they remained a local curiosity until the 1830s when, after intensive study, it was concluded that they had been made by gigantic ancient birds. They were not recognized as belonging to dinosaurs until the 1870s.

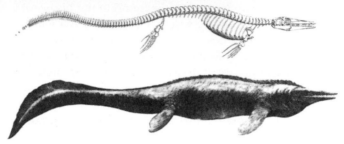

Cuvier's *Mosasaurus*
The jaws of the
Maastricht
Mosasaurus *(top)*
were the first
indication of gigantic
fossil reptiles. This
large, sea-going lizard
(below) is not closely
related to dinosaurs.

terns created by God, and also that the shapes of bones indicated their function in life – for example whether the animal ran, walked, flew or swam. With this in mind Cuvier studied and dissected large numbers of animals. He that realized that by understanding the bones of living animals he could develop a method for understanding how fossil animals, which might be known only as poorly preserved or fragmentary pieces of bone, might have looked in life. This technique is called comparative anatomy, because it requires comparisons to be made between types of animal in order to understand why they have that form.

Cuvier soon gained a worldwide reputation as the most knowledge-able of all anatomists of his day. He put his skill to excellent use in two particularly important, and at the time controversial, ways.

By reviewing fossils he collected from various parts of the world (he concentrated particularly on elephants) he was able to prove that some animals had become extinct in the history of the Earth. This was a revolutionary notion at the time because it directly challenged religious views which stated that God populated the world with his creatures and would not have let any of them disappear. However, Cuvier's anatomical arguments gradually prevailed, thereby prompt-ing a great deal of interest in the fossil record.

The other crucial demonstration, which built on the rising tide of interest in fossil creatures, came through a Napoleonic spoil of war. In 1795 the French Republican army occupied the southern Dutch town of Maastricht and, having sacked it, brought back an unusual trophy. It was a slab of chalk which had been found in a local quarry a few years earlier; and it contained the jaws and skull bones of a gigantic creature. The fossil was taken to the Jardin des Plantes, where Cuvier was able to study it. He was able to identify the creature as an extinct lizard related to the large monitor lizards of the tropics – but this was no ordinary lizard; it was a gigantic one, whose head alone was 4 ft (1.2 m) long! Later the creature was to be named *Mosasaurus* (literally "Meuse lizard") by the English geologist the Revd W.D. Conybeare, for the area from which it came. In life it was a gigantic, seagoing lizard with a long, powerful tail and paddle-shaped legs for swimming.

With these two critical discoveries Cuvier showed that fossils could be interesting not only because they were different from animals living today, but also because some truly stupendous creatures were out there to be found. Cuvier's views on such matters became extremely well known after he published a massive series of books, *Recherches sur les Ossemens fossiles* (Studies of Fossil Bones), which became the standard work for aspiring paleontologists and comparative anatomists.

William Buckland and Megalosaurus

William Buckland (1784-1856) was an extraordinary man of many talents. Not only was he a distinguished churchman – later becoming Dean of Westminster – but he was the first Professor of Geology at Oxford.

Some time before 1818 a collection of large fossil bones and jaws with teeth was made at a slate quarry in the village of Stonesfield in north Oxfordshire, about twelve miles (19 km) north of Oxford; these fossils were passed to Buckland for identification. In 1817 and 1818 Cuvier was paying his first visits to England, and he travelled to Oxford to meet Buckland and examine the new fossils. It was apparent to Cuvier that Buckland had the remains of a new and unknown giant fossil reptile, and that it was rather similar to the fossils from Normandy that he had described as crocodiles. Buckland published a report on the find in 1824 in the *Transactions of the Geological Society of London* naming it *Megalosaurus* ("big-lizard").

Megalosaurus was by no means complete. Its remains consisted of a lower jaw with large, blade-like teeth; some vertebrae; shoulder bones; part of the hip; and several bones of the hind leg. But even that was far more than earlier workers had had to examine. Buckland suggested that *Megalosaurus* was a giant extinct predatory lizard, in much the same way as Cuvier had proposed for the Maastricht fossil skull. Cuvier had also advised Buckland that this animal, judged by the size of the hind leg bones, may have been more than 40 ft (12 m) long and had the bulk of an elephant 7 ft (2 m) high. Buckland may

The jaws and teeth of Megalosaurus *were described in detail by William Buckland. These remains were quite unlike Cuvier's mosasaur and led him to propose that this was perhaps an elephant-sized, land-living, predatory reptile. The large, curved tooth projecting from the jaw has fine, saw-toothed edges, ideal for cutting through flesh.*

not have realized that he was dealing with a dinosaur, but it was clear that the animal was quite unlike living lizards.

Buckland also referred in his scientific article to another English fossil collector and scientist of the time, Dr Gideon Algernon Mantell. Mantell had also discovered megalosaur bones – some even larger than Buckland's as well as some other puzzling fossils.

Gideon Mantell and Iguanodon

Gideon Algernon Mantell (1790-1852), a doctor living in Lewes on the south coast of England, was an enthusiastic geologist, spending much of his free time surveying the neighbouring countryside and collecting the abundant fossils of the region. In 1822 he published a book, *The Fossils of the South Downs*. This included accounts of some unusual fossil teeth, angular sided, some of them heavily ground down, which had been discovered in the quarries of Tilgate Forest. In trying to identify these teeth he consulted all the known experts on fossils in Britain. Buckland believed them to be the front teeth of relatively recent mammals, whose remains had been mixed up in older rocks. Unsatisfied with any of these identifications, Mantell sent some examples to Cuvier in June of 1824. Cuvier replied within the month and strongly supported Mantell, eliminating any possibility of the teeth belonging to fish, and suggested that they were from a large herbivorous reptile, previously unknown. Cuvier published a brief report on these in his 1824 volume of *Ossemens fossiles*, pointing out their superficial similarity to the heavily worn front teeth (incisors) of a large fish or of a mammal such as a rhinoceros, but concluding that they were likely to be the remains of a reptile, and perhaps one of the most extraordinary yet discovered.

70

Later that year Mantell had some luck. On visiting the Museum of the Royal College of Surgeons at Lincoln's Inn Fields in London, he was shown the newly prepared skeleton of an iguana. Although the teeth of this Caribbean lizard were minute by comparison with the fossil teeth, they were quite similar in shape, and it was a herbivorous reptile. In this report on the fossils in the *Philosophical Transactions of the Royal Society* for February 10, 1825, he called the creature *Iguanodon* ("iguana tooth"), which had been suggested to him by the Reverend William Conybeare (who had also coined the names *Megalosaurus* and *Mosasaurus*). Comparing the fossil teeth with those of the living iguana, Mantell concluded that *Iguanodon* may have been even larger than *Megalosaurus*, suggesting a length of 60 ft (18 m).

Curiously, William Smith, who created the first geological maps of England and Wales (see page 31), had already collected from Mantell's hunting grounds. In 1978 a collection of fossils made by Smith at Cuckfield in 1809 was rediscovered by Dr Alan Charig in the collections of the Institute of Geological Sciences – now part of the Natural History Museum, London. Among the many poor and fragmentary pieces there is one very distinctive part of the lower leg of *Iguanodon*. William Smith was entirely unaware of the significance of the discovery, but this seems to be the earliest authenticated discovery of this dinosaur.

In 1833 Mantell found another large reptile in the same quarries in Tilgate Forest. This one, which he named *Hylaeosaurus* ("forest lizard"), consisted of the front half of an animal considerably smaller than *Iguanodon*, with evidence of large pointed spines running down its back. Then a year later, a partial skeleton of *Iguanodon* was discovered by workers in a quarry at Maidstone in Kent. It was bought for Mantell by a group of friends, for the sum of £25.

MANTELL'S RECONSTRUCTION OF *IGUANODON*
This partial skeleton of Iguanodon *(left) preserved in a slab of rock, affectionately known as the "Mantel-piece," formed the basis for Mantell's sketch (right) in the mid 1830s.*

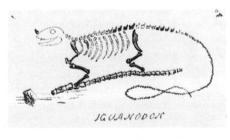

Owen's reconstruction of Megalosaurus *was based on very meagre evidence indeed: the lower jaw, some ribs, a toe bone, the hips and thigh bone. Yet from this he managed to create this massive bear-like structure.*

The new specimen allowed Mantell to attempt the first ever reconstruction of *Iguanodon* with the aid of William Clift, Curator of the Museum of the Royal College of Surgeons. The result clearly shows the strongly lizard-like proportions which this creature was believed to possess. The spike on the nose in this illustration was put there by Mantell after he found a curious conical bone at Tilgate, which seemed similar in shape to the nose horn found on some iguanas.

Richard Owen and the Dinosauria

In 1827, when Mantell was most busy with his early discoveries of *Iguanodon*, the twenty-three-year-old Richard Owen was appointed assistant to William Clift at the Museum of the Royal College of Surgeons in London. One of the tasks which came with the job was to dissect and describe the various animals which died at the London Zoo (or Zoological Gardens, as it was then known). Owen was extremely hardworking and rapidly became a skilled comparative anatomist.

In 1837 Owen became the Hunterian Professor of Comparative Anatomy and began to publish a large number of scientific articles on fossil mammals and reptiles. At about this time he also conceived the idea of undertaking a review of all known British fossil reptiles, and made a formal proposal to the British Association for the Advancement of Science in 1838. They agreed to sponsor his work, and over the next few years he travelled the length and breadth of the country looking at and describing fossil reptiles. In 1839 Owen presented his report on the Enaliosauria or "sea lizards" to the British Association. On Friday 30, 1841 the second and final part was read at the eleventh meeting of the British Association.

By far the greatest innovation of Owen's work was made while he was writing his report of the meeting. In this, he suddenly realized that the giant reptiles which had been described as *Megalosaurus*, *Iguanodon* and *Hylaeosaurus*, formed a natural group and required a name in recognition. The name Owen chose was "Dinosauria" (the "terrible" or "fearfully great lizards") which first appeared in print in 1842. Owen saw in them not lizard-like features, as had been widely assumed (as in Mantell's early sketch of *Iguanodon*) but an unexpected mixture of features. These were: double-headed ribs in the chest, like those of crocodiles; the great height, breadth and sculpturing of the spines of the backbone, which were unusual in their own right; five fused sacral bones – that is, lower vertebrae – attached to the pelvis, as in mammals; long, hollow limb bones, with prominent processes (projections) for muscle attachment, indicating that the creatures moved on land, like mammals; toe bones which, apart from their sharp claws, strongly resembled those of heavy living mammals such as rhinoceros, elephant and hippopotamus, again a mammal-like feature; complicated shoulder bones, as in lizards; and teeth intermediate in type; *Megalosaurus* teeth were embedded in sockets in the jaw, rather like those of a crocodile; those of *Iguanodon* and *Hylaeosaurus* were more like the teeth of lizards.

With the wisdom of hindsight it is possible to see that Owen missed several other dinosaurs which were in his report, and was simply naming a group of reptiles which Cuvier had recognized as "extraordinary" seventeen years earlier. In 1824 Cuvier had advised Buckland of the similarity in build and proportions between *Megalosaurus* and a medium-sized elephant. In the same year, as we have seen, Cuvier had offered the provocative suggestion to Mantell, which was published in Mantell's article of 1825, that the teeth and bones he had discovered at Tilgate seemed to represent the remains of a gigantic, herbivorous reptile comparable to the large herbivorous mammals of the present day.

Owen took these suggestions a step further by providing an official name for these elephantine lizards and in conjuring up an image for them, but in so doing he was gambling on his scientific reputation and the comparative method developed by Cuvier. It was quite a risk, which paid off handsomely at first; but in the end it was not fully vindicated. The risks were clear enough. As we have seen, the three dinosaurs that he selected were very poorly known from only a few bones and teeth. No one could be sure of their appearance in life; but Owen was sufficiently sure of the comparative method to risk it anyway.

Owen also seems to have had an ulterior motive for creating dinosaurs, which may explain this daring stance in 1842. This was to counter a very strong line of argument, then current among some French and British anatomists, concerning the evolution of animal life (or "transmutation", as it was then known). One particular line of

73

reasoning was underpinned by the observation from fossils that life had become progressively more complex with time – the so-called "progressionist" movement. Owen fundamentally disagreed with this philosophy, and believed that dinosaurs would enable him to prove his point. He argued that dinosaurs were in their anatomy and physiology far superior to the reptiles that we see today. Modern reptiles, he argued, are degenerate forms compared with the magnificent dinosaurian reptiles of the Mesozoic Era.

You can see how far Owen was prepared to go in order to make his point, in the summary to his 1842 report. He speculated that the atmosphere in the Mesozoic Era was lower in oxygen and more suited to the less energetic reptiles than to the more energetic mammals and birds. However, he suggested that dinosaurs may have led more energetic lives than most reptiles, because they had a four-chambered heart more like that of mammals and birds, and, he said, were likely "...from their superior adaptation to terrestrial life, to have enjoyed the function of such a highly organized centre of circulation [the heart] in a degree more nearly approaching that which now characterizes the warm-blooded vertebrate [that is, mammals and birds]."

Thus Owen ended his report, and with considerable courage and great farsightedness anticipated the physiological arguments which have dominated dinosaur studies for the past twenty years: whether dinosaurs were "warm-blooded" or "cold-blooded." He also inadvertently provided one of the earliest theories for the extinction of the dinosaurs: which was that the level of oxygen in the atmosphere rose (or the atmosphere was in some other way "invigorated") until conditions proved intolerable to such reptiles.

Owen's opportunity to give dinosaurs real and lasting substance came unexpectedly in 1852 when, as recounted earlier (page 13) he was given the chance to collaborate with Benjamin Waterhouse Hawkins in making life-size dinosaur models for the Crystal Palace park at Sydenham.

As a visitor can see today, the models are most imposing – yet they are anything but accurate. The large, heavy-limbed monsters are rather reminiscent of modern pachydermal (thick-skinned) mammals such as the rhinoceros, except that they have heavily scaled skin and rather long, reptilian tails. This is particularly the case with Mantell's *Iguanodon* – of which there are two models, each equipped with a prominent horn on the tip of its nose. *Megalosaurus* seems to resemble a very large, long-snouted bear; while *Hylaeosaurus* seems to be more like a slender version of the *Iguanodon*, with a spine-encrusted back and minus the infamous nose horn.

It is easy to poke fun at Owen's dinosaurs now, as the American paleontologist O.C. Marsh did when visiting England in 1895, three

Hawkin's Crystal Palace workshop: Iguanodon *(centre), rear view of the spiky* Hylaeosaurus *(right), a gigantic frog-like amphibian and a tortoise-like dicynodont (front)*

years after Owen's death. "So far as I can judge, there is nothing like unto them in the heavens, or on the earth, or in the waters under the earth. We now know from good evidence that both *Megalosaurus* and *Iguanodon* were bipedal, and to represent them as creeping, except in their extreme youth, would be almost as incongruous as to do this by the genus *Homo*." But they were a brilliant innovation in their time, based as they were on extremely poor fossil material, and it is very unlikely that anyone, especially Marsh, could have come up with anything better at the time.

In later years Owen, who was probably at the height of his intellectual power in the mid-1850s, saw his conception of the dinosaurs gradually superseded by newer discoveries of more complete skeletons of dinosaurs. These mostly showed that his comparative anatomy was only as good as the material which was used for comparison. In the case of dinosaurs the body plans were completely different from anything that he could have reasonably guessed – which is why Owen's dinosaurs look so antiquated.

DINOSAURS AFTER OWEN

A short while after Owen's models had been completed, dinosaur remains began to be discovered in America. Footprints had been found many years earlier, but the first discoveries which led to proper scientific descriptions were made in 1855 by Ferdinand Vandiveer Hayden, during the course of an expedition to what is now eastern Montana. In an area near the confluence of the Judith and Missouri Rivers a number of unusual teeth were found in Cretaceous rocks. These were passed on to Joseph Leidy (1823-1897), who was professor of anatomy at the University of Pennsylvania.

Leidy published short descriptions of these teeth the following year. Two of the teeth Leidy considered to be lizard-like, and he named them *Palaeoscincus* ("ancient skink") and *Troödon* ("wounding tooth") – both of these have since turned out to belong to dinosaurs as have several other teeth found at the site. Thus, even though Leidy only had the evidence of a few teeth, he was able to draw comparison with the European finds which had included dinosaurs.

The Central Park Commissioners had very grand plans for their Paleozoic Museum. An artist's impression shows that it was intended to resemble Paxton's Crystal Palace except that the prehistoric animals were inside rather than outside. It was to be a place celebrating the ancient life that once inhabited the North American continent.

Charles Knight's painting of Laelaps *catches a very different mood when compared to Owen's elephantine dinosaurs. It captures the kangaroo-like features of the dinosaur that Cope had noted and paved the way for more sprightly images of dinosaurs.*

Within two years the evidence of dinosaurs in North America was far better than that which had accumulated in thirty years of searching in Europe. The discovery came to light very close to Philadelphia and was made by William Parker Foulke. In 1858 a partial skeleton was dug up at Haddonfield, New Jersey, and handed over to Leidy. Leidy quickly described and named the new specimen *Hadrosaurus foulkii* ("Foulke's heavy lizard"), and pointed out that it was again rather similar to *Iguanodon*. However, this specimen included nine teeth, a part of the jaw, many vertebrae – and, most importantly, the main bones of the limbs, which indicated that the hind limbs were much longer and stronger than the fore limbs. Leidy therefore suggested that these animals may have walked on their hind legs alone, giving them rather a kangaroo-like stance, in complete contrast to the elephantine reptiles imagined by Owen.

Ten years later Leidy's *Hadrosaurus* was destined to come to life in the hands of Benjamin Waterhouse Hawkins – the sculptor of Owen's dinosaurs – for the Paleozoic Museum planned for Central Park, New York. Alongside two models of *Hadrosaurus*, another carnivorous

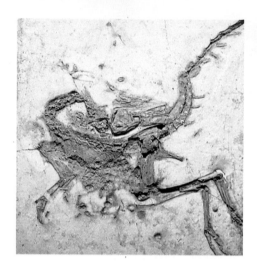

The discovery of this exquisitely preserved skeleton of the bird-like Comsognathus *focused attention on the structure of dinosaurs.*

dinosaur was modelled; this was *Laelaps* (named for the hunting dog in Greek mythology which was turned to stone in mid-leap). This was also known from a partial skeleton which had been discovered in 1866 by a precociously talented student of Leidy's, Edward Drinker Cope (1840-1897). Cope had also been able to demonstrate that his creature had an even greater difference in the size of its fore and hind limbs.

In just a few short years ideas about the anatomy and form of dinosaurs were changed dramatically by the American discoveries. Dinosaurs, however, were still a rather confusing lot to the experts of the time. Meanwhile new discoveries were being made in Europe. In 1861 the famous early fossil bird *Archaeopteryx* ("ancient feather" – see page 91) was discovered in limestone quarries in Bavaria. In the same year another small (2ft, 60 cm long) complete fossil reptile skeleton was discovered in the same area of southern Germany and was named *Compsognathus* ("pretty paw"). Both Cope and the English anatomist Thomas Henry Huxley (1825-1895) realized in the early 1960s that *Compsognathus* was not just any reptile, but a tiny dinosaur. It was lightly built, with long, bird-like legs and feet and a similar posture to that of *Hadrosaurus* and *Laelaps*. This suggested that there were greater similarities between birds and dinosaurs than between mammals and dinosaurs, as Owen believed.

At the same time Owen was still describing new dinosaurs in England and, rather unexpectedly, these seemed to conform to Owen's ideas. Two in particular which he described were *Omosaurus*, from the Late Jurassic, a stegosaur; and *Scelidosaurus* from the Early Jurassic, an early ankylosaur. Both of these were heavily armored types, most unlike previously described forms, and both showed clear evidence of having walked on four rather than two feet. It was clear that more evidence was needed to break the deadlock between Owen's theory and the rival theories of Huxley and Cope.

Dinosaur riches in Belgium

In April 1878, miners were excavating a coal seam in the village of Bernissart in southwest Belgium when they found what proved to be fossil bones. At once a cable was sent to scientists at the Royal Museum of Natural History in Brussels. Fortunately some teeth were preserved with the bones, proving beyond doubt that what had been discovered were the remains of one of Owen's first dinosaurs, *Iguanodon*.

The find took many years to sort and prepare, and yielded a huge cross-section of Early Cretaceous life: fish, plants, tortoises, crocodiles, insects, and of course dinosaurs. Parts of complete skeletons of thirty-nine *Iguanodon* were discovered, and fragments of a solitary carnivorous dinosaur, *Megalosaurus dunkeri*.

In 1882 Louis Dollo (1857-1931), was appointed as a museum assistant and given the job of describing the fossil reptiles. He devoted much of the rest of his life to his work. In a long series of scientific articles Dollo showed that it was Leidy, Cope and Huxley who had been correct about the general shape and postures of these animals. The large number of skeletons of *Iguanodon* proved beyond any doubt that the animals had longer and stronger hind legs, and a long muscular tail. In addition the feet were very bird-like, with three long forwardly directed toes, and to further confirm the similarity to birds, the bones of the pelvis had a distinctly bird-like arrangement.

Dollo is still remembered today by biologists of all kinds, for his name is given to an evolutionary proposition which he developed, known as "Dollo's law of the irreversibility of evolution." This law is important in considering the evolution of birds (see page 196). He was also able to clear up a longstanding puzzle – where to put the conical bone which Mantell, Clift and later Owen had put on the nose of *Iguanodon*. Just a year after the Crystal Palace models of *Iguanodon* were finished Owen had second thoughts about the supposed rhinoceros-style "nose-horn" and suggested that it might be a sharp claw on the hind foot. Some time later another specimen was discovered which was attached to the forearm bones of *Iguanodon*. Dollo showed that this bone was the greatly enlarged thumb claw of the hand – a savage weapon.

The American dinosaur rush

Just a year before the dramatic discoveries at Bernissart, richer and more varied deposits were found in Colorado. They were discovered

Arthur Lakes painted many scenes from the dinosaur collecting grounds in Colorado and Wyoming. Here paleontologists expose dinosaur bones from the Morrison Formation at Como Bluff, Wyoming - these brontosaur remains are now on display in the USA. By sending his fossils to the paleontologists Cope and Marsh, who were arch rivals, Lakes unwittingly helped to precipitate a dinosaur rush as both men competed to make important new dinosaur finds.

independently and apparently by chance, by two school teachers, Arthur Lakes and O. W. Lucas.

Lakes found his fossils near Morrison, Colorado, in the foothills of the Rocky Mountains. At this time there were two particularly well known paleontologists in America. One was Edward Drinker Cope at Philadelphia, whose work on *Laelaps* has already been mentioned (page 123), and who had described some of the dinosaurs of New Jersey in 1860s, and later led an expedition through the Sioux Indian nation territory to the Cretaceous deposits in the Judith River area of Montana in 1876, where he collected one of the first horned (ceratopian) dinosaurs, *Monoclonius*, as well as some hadrosaurs. Othniel Charles Marsh at Yale College (later University) had also described some dinosaur remains from New Jersey (*Hadrosaurus minor*) in 1870, as well as *Claosaurus* from Kansas. For this reason Lakes sent some of these large bones to Marsh, who was then Professor of Paleontology in the Peabody Museum at Yale. The remainder of his collection he sent to Cope.

However, Cope and Marsh were bitter enemies. The source of the feud between the two men seems to date from 1870. In 1868 Cope had described a new marine fossil reptile known as *Elasmosaurus* ("ribbon lizard") which, he suggested, was very unusual in the structure of its backbone. Marsh came to examine Cope's new reptile in 1870 and, after looking at it, suggested that Cope had been mistaken and had reconstructed the animal with its head placed on the end of its tail! Unfortunately for Cope, Marsh was correct. Cope was devastated.

As soon as Marsh had examined the bones sent by Lakes, he hired Lakes, instructing him to keep his discovery secret – presumably an attempt to avoid competition from Cope. Cope, however, had already received his consignment of bones, and was busy describing them when he received a message from Lakes asking him to pass the bones on to Marsh. It is not hard to imagine the frustration that Cope must have felt in this situation. The advantage which Marsh enjoyed, however, was not long-lived. The other teacher, O. W. Lucas, had found some gigantic bones at Garden Park near Canyon City, Colorado, in another part of the same geological formation in which Lakes' discoveries had been made. He sent these specimens directly to Cope. So began an intense rivalry and a rush to publish information on the new discoveries. The bones found at Canyon City were larger and more complete than the ones from Morrison, so it appeared that Cope would gain more than Marsh. But Cope's lead did not last for long either. Later in the same summer of 1877 new discoveries were made at Como Bluff, Wyoming. This time Marsh was first on the scene. Vast numbers of bones were collected and shipped back to him at Yale over the next twelve years.

The discoveries made by Marsh and Cope in Colorado and Wyoming revealed a host of ancient giants from the Upper Jurassic, including *Allosaurus, Ceratosaurus, Camarasaurus, Brontosaurus* (later called *Apatosaurus*), *Amphicoelias, Diplodocus, Stegosaurus,* and *Camptosaurus.* But in the late 1880s attention shifted to the Cretaceous rocks. Cope had started the search here in 1876, but abandoned it in the heat of the dinosaur rush to Colorado and Wyoming. With the assistance of John Bell Hatcher, small carnivorous dinosaurs (*Ornithomimus*) and armored dinosaurs (*Nodosaurus*), as well as large quantities of horned dinosaurs, were collected from the Judith River area – including large numbers of skulls and skeletons of the famous *Triceratops.* These dinosaurs passed to Marsh and brought him more fame in the late 1880s and 1890s.

After the great rush

Following the deaths of Cope in 1897 and Marsh in 1899 the frenetic search for dinosaurs abated, and gave way to a time of slower and more careful excavation. In 1897 new expeditions were made to Como Bluff by the American Museum of Natural History under the guidance of Henry Fairfield Osborn. Osborn was anxious to obtain collections as good as those at Yale. Two skeletons were found in the first season, but then in 1898 a new locality was discovered close by which proved to be incredibly rich. Hillsides were literally strewn with dinosaur bones, so abundant that a shepherd had built a small shelter out of them. The area became known as Bone Cabin Quarry. Over the next seven years it produced fabulous quantities of fossil bone, which formed the basis for a new dinosaur hall at the museum.

Andrew Carnegie's dinosaur

The American Museum was not the only museum interested in building up collections of fossils, and especially dinosaurs. The Carnegie Museum in Pittsburgh, set up by the great industrialist and philanthropist Andrew Carnegie, was also anxious to obtain dinosaurs; in fact it was one of Carnegie's pet ambitions. Newspaper articles publicizing the new dinosaur finds from the American Midwest attracted Carnegie's attention, and he sent word to the Director of his museum, W. J. Holland, that he wished to find a giant dinosaur for his museum. With Carnegie's funding, Holland assembled his own crew of fossil collectors and went to Wyoming in 1899.

Here a cast of the Carnegie specimen of Diplodocus *is formally presented to the Natural History Museum, London, in 1905 in the presence of King Edward VII. Andrew Carnegie had offered to make the cast (which took two years to complete) after the King had visited Carnegie's Scottish home and admired a painting of the skeleton. At the time, this* Diplodocus, *displayed at the Carnegie Museum in Pittsburgh, was the biggest mounted skeleton in the world.*

Within two months his team made a spectacular discovery at Sheep Creek, near Medicine Bow: the skeletons of two *Diplodocus*. Neither skeleton was complete but, by patching the two together back at Pittburgh, Holland was able to present Andrew Carnegie with his own personal dinosaur. At the time it was the biggest mounted skeleton in the world, and to cap it all it was named *Diplodocus carnegiei*. Carnegie, delighted with his dinosaur, not only put it on display in Pittsburgh but commissioned paintings of the creature. One of these was sent to Carnegie's home in Scotland, where it was seen by Edward VII, the King of England, on a visit in 1903. The King was much impressed with the painting and wondered whether it might be possible for Carnegie to obtain such a dinosaur for Britain. Carnegie offered to have a cast made of the *Diplodocus*, and for the next two years a team of Italian plasterers was employed to mold and cast the entire skeleton of the dinosaur. In 1905 the complete replica was assembled in the Natural History Museum in London and officially opened at a royal occasion. The event was so successful that Carnegie had further casts made and sent to the museums of many capital cities around the world.

A second big break for the Carnegie excavators came in Utah in 1909, when Earl Douglass of the museum, accompanied by George Goodrich, a local Mormon elder, went to the foothills of the Uinta Mountains, near Vernal in Utah, in search of dinosaurs. Earlier survey reports had suggested that bones were to be found, but these reports were rather vague. In the end Douglass found, in a wall of tilted rock strata, a row of eight dinosaur tail bones in connection. Dr Holland came out from Pittsburgh to see the find, and as the line of the row of vertebrae was followed and excavated, more and more of the skeleton was revealed. In the end a virtually complete gigantic dinosaur was exposed, and was to be called *Apatosaurus louisae* in honour of the wife of Andrew Carnegie. But for this particular site, that was just the beginning. From 1909 until 1923 crews from the Carnegie Museum regularly visited it, finding more *Diplodocus*, as well as *Apatosaurus, Camarasaurus, Stegosaurus,* and *Allosaurus.*

The Canadian dinosaur rush

Blackfoot Indians knew of the existence of dinosaur bones in the valley of the Red Deer River many centuries before the white man discovered these remains. They were thought to be the bones of the ancestor of the buffalo, and offerings were made to the spirit world to give braves good fortune when hunting. Scientific exploration and research began in the 1870s, with the beginning of the geographical surveys of the Canadian borders. In 1884 Joseph Burr Tyrrell made the first important discovery, the skull of a carnivorous dinosaur in the Late Cretaceous rocks of the valley of Red Deer River. He sent the fossils to Ottawa for study by the Canadian Geological Survey. However, at the time there was no one in Canada capable of identifying them so they were sent to Cope in Philadelphia, who identified the remains as those of a carnivorous dinosaur. This was eventually given the name *Albertosaurus* – celebrating its discovery in what was later called the Province of Alberta.

Now the search for the dinosaurs in Canada began in earnest. Lawrence Lamb of the Canadian Geological Survey discovered a number of fossils, and described many of them; but the collecting techniques which he used did not produce much good material. The badland terrain was very difficult both to explore and to work, compared with the fossil fields of America. Then in 1910, a collector for the American Museum of Natural History in New York, Barnum Brown, had a simple but brilliant idea. In the previous year, following the rumor of rich dinosaur fossils in Alberta, Brown had conducted an exploratory expedition in the Red Deer River. He returned fired with enthusiasm for a major expedition. So for the 1910 season he decided to built a large raft which could carry all the equipment and act as a mobile camp. This they could pole down river to likely spots, moor and then explore the exposures of rock for any bones. They would load their specimens on board, and the expedition moved on.

Each season the raft would end up laden with fossil treasures, including several complete dinosaurs, which Brown shipped back to New York to form the basis for some of the marvelous displays of dinosaurs in the museum's Cretaceous Hall. Brown had been supported at first by the Canadian Geological Survey but, once they realized the riches that he was recovering, they hired their own team to collect dinosaurs. Charles Hazelius Sternberg (who had collected with Cope many years previously) and his sons Charles Jr, George, and Levi built a raft and copied Brown's technique, also to great effect.

The African expeditions

The year 1907 saw the discovery of gigantic dinosaur remains in Late Jurassic rocks in Tendaguru, in what was then German East Africa and is now Tanzania. The dinosaurs were excavated between 1908 and 1912, at enormous cost and effort, by local workers under the supervision of Edwin Hennig and Werner Janensch of the Berlin Museum for Natural History. Transporting the finds to Berlin from a remote roadless part of Africa presented a logistic nightmare. All the bones, which amounted to about 250 tons when packaged up, were carried on the heads of hired porters, or strung between them on poles. In this way they went some 40 miles (65 km) to the port of Lini, from where they were shipped back to Berlin.

The human and financial efforts were well worth it. The expedition was wonderfully successful and revealed a whole fauna of dinosaurs, including the spiky stegosaur *Kentrosaurus*, the slender theropod *Elaphrosaurus*, and *Diplodocus*-like *Dicraeosaurus*, and the giant sauropod *Brachiosaurus*. The latter dinosaur, which stands some 40 ft (12m) tall, and is over 74 ft (22.5m) in length, is the crowning glory of the Berlin Museum. It is the biggest mounted dinosaur skeleton in the world, dwarfing the cast of Andrew Carnegie's *Diplodocus* which stands by its side.

More Dinosaurs from Europe

In addition to the discoveries in her colonies, Germany also had some exceptional dinosaurs closer to home. Numerous bones of dinosaurs dating from the end of the Triassic had been found in southern Germany since the late 1830s – some of the larger bones were described and given the name *Plateosaurus* by Hermann von Meyer in 1837. Many of these were reviewed and studied by Friedrich Freiherr von Huene (1875-1969) a distinguished German paleontologist, who was based at the University of Tübingen for most of his working life. While much of his early work was concerned with the classification of dinosaurs and re-studying other people's work, in 1921 von Huene was involved in a marvelously rich discovery of dinosaurs at a quarry near Trossingen just 30 miles (48 km) south of Tübingen. Much as in the case of the *Iguanodon*, some intriguing fragments had been discovered many years earlier, to be followed by incredible riches. The quarry at Trossingen yielded large numbers of complete skeletons of *Plateosaurus*.

The Central Asiatic expeditions

The early 1920s saw the American Museum not only aiding von Huene, but hatching an ambitious plan. Osborn, who had been involved in dinosaur discoveries in Wyoming at the turn of the century, proposed an expedition to Mongolia to discover evidence of the origin of the human race. The prevailing view of the time was that this had occurred in Central Asia, and Osborn intended to test the theory. Despite the remoteness of their destination, and all the logistical and political difficulties, the expedition finally arrived in Mongolia in 1922. Some mammal remains were discovered, but no significant human relics. Instead the expedition stumbled upon curi-

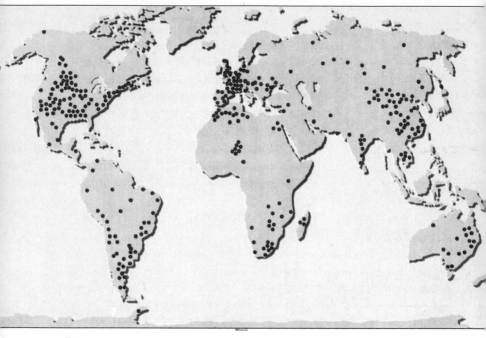

DINOSAUR DISTRIBUTION
Dinosaurs have now been found on every continent on the globe. Within the last few years remains have even been discovered in the frozen wastes of Antarctica by teams from Britain and Argentina. The traditional areas for dinosaurs, western Europe and North America, are still producing important new finds, but exciting discoveries are being made particularly in Mongolia, China, South America, Africa and Australia.

ous beaked dinosaurs (*Protoceratops, Psittacosaurus*), strange armored dinosaurs (*Pinacosaurus*), and peculiar carnivorous dinosaurs (*Saurornithoides, Velociraptor,* and *Oviraptor*). They had found a rich Late Cretaceous dinosaur graveyard, and even dinosaur eggs and nests – the first of any dinosaur ever to be discovered.

Chinese dinosaurs

The huge land area of China is also very rich in dinosaur fossils. Early finds (other than the traditional "dragon's teeth") date back to around 1900, when a Russian general collected bones which had been found by fishermen in northern China. Following this several expeditions visited China: joint Sino-French and Sino-Swedish enterprises, and the Osborn expeditions on their way to and from Mongolia. However, from 1933 onward the Chinese took sole responsibility for expedition work. Since that time China has steadily produced a series of unusual dinosaur discoveries.

Dinosaurs from all over the world

These are by no means the only places where dinosaurs have been found. South America has yielded large quantities of remains, including the amazing large-horned theropod *Carnotaurus* and the armored sauropod *Saltasaurus*, both found in Argentina. Late Cretaceous theropods have also come from the Deccan region of India. In Australia a large theropod, *Rhoetosaurus*, was discovered in the 1920s near Brisbane, and more recently various ornithopods have been found. Even Antarctica has yielded several armored dinosaurs and an ornithopod; and at the other end of the world, scientists from the University of California at Berkeley have found remains in Alaska. So it is true to say that there were dinosaurs in every corner of the world.

ABOVE: *The dinosaurs of China have proved to be consistently interesting. This is a group of medium-sized sauropods* Omeisaurus tianfuensis *moving through a Late Jurassic forest. Whether the tail ended in a stiff rod is uncertain.* RIGHT: *Some fossils still present perplexing problems.* Hallucigenia *lived on the sea bed 535 million years ago. We now know that it walked on elongate tentacles, defending itself with the prominent spines that rose along its back. However, it is unclear which end is the front, though it is likely to be at the bottom of the picture, as shown.*

CHAPTER FOUR

Bringing Dinosaurs Back to Life

Paleontologists do not just find and excavate dinosaurs. They return their specimens to museums and organize their cleaning and mounting in galleries, which in itself can be a very expensive and enormously time-consuming process. But they are also keen to learn about the nature of the animals they discover. The difficulty of this task depends very much on how closely related the fossils are to living creatures. For example, paleontological discoveries of fossil elephants such as mammoths are greatly aided by our knowledge of living elephants; the anatomy of the living species can be used to help to understand the similarities and differences in the fossil types with some considerable accuracy. However, the farther you go back into the fossil record, the less similar to modern, living creatures are the fossils that are found. To take an extreme example, samples of fossils of creatures from the Middle Cambrian Burgess Shale of British Columbia (535 mya) include many soft-bodied creatures which lived in or on an ancient sea floor; many of these, such as the aptly named *Hallucigenia*, are a total mystery since they bear not the least similarity to today's species.

Dinosaurs are not totally alien: at least we can sort our their basic anatomy with some confidence. But they are not closely comparable to living animals, and it is this curious mixture of known and unknowable features which can make them such interesting, and at the time such frustrating creatures to work on. The story of the rise of our present understanding of these amazing creatures is one with a surprising number of twists and turns. It is not a steady and consistent improvement in our knowledge. Instead, progress took the form of a series of lurches – many of them were certainly not forward.

Gigantic lizards

The struggle to understand the nature and way of life of dinosaurs dates back to the 1820s and the pioneering work of Buckland, Mantell and Cuvier. From the time of the earliest studies it was evident that dinosaurs were interesting and unusual creatures. To Buckland and Mantell their discoveries were little more than the remains of gigantic lizards. Cuvier, however, seems to have grasped the unusual nature of these reptiles from the beginning. It was he who drew the comparisons with modern large mammals such as the elephant, and put forward the idea that these were new and previously unknown types of reptile. His words and thoughts were remarkably far-sighted when you consider how meager his evidence was. The poor quality of this early material had a lot to do with the slowness of other early anatomists to appreciate how different and unusual dinosaurs were.

Elephantine reptiles

The giant lizards of the 1820s and 1830s gave way to elephantine reptiles when Professor Richard Owen grasped the thread of understanding offered by Cuvier. Providing the creatures with a name – Dinosauria – and identifying certain features which he believed that they possessed in common (strong hips, upright legs, and short feet) gave them an identity which began to focus interest on them as a group, rather than simply as odd, but interesting, individual fossil reptiles. Owen took the discussion and theorizing much further. Perhaps Owen's reasoning can be doubted, because his motives seem to have been to disprove "progressionism." He certainly went to great lengths in his discussion of the possible biology of these animals. Not content with demonstrating certain structural similarities in the hips, legs and feet to those of large living mammals, he went on to speculate upon the system of the chest, heart, and circulatory system. The reconstructed models of dinosaurs for the Crystal Palace brought his vision of these animals to life in the eyes of the Victorian public and scientific community alike.

In the late 1860s Professor Owen's model of dinosaurs as large, elephantine lizards became subject to severe criticism by other scientists. Foremost among these was Thomas Huxley who was able to describe small, extremely agile dinosaurs such as Hypsilophodon, *which was then newly discovered in southern England.*

Of Kangaroos and birds

Yet Owen's vision of dinosaurs immortalized in concrete in 1854 was based on very meager evidence and was obviously open to errors of interpretation. Within a few years of the completion of the Crystal Palace dinosaurs, new evidence was to emerge from the New World which would challenge Owen's elephantine reptiles.

Leidy's description in 1858 of the partial remains of the dinosaur *Hadrosaurus* from New Jersey provided the first clear evidence of the posture of dinosaurs. The teeth of this creature were rather similar to those of the British *Iguanodon*, but the front and hind limb bones were very different in length, with rather short fore limbs and long, heavier hind limbs, like those of a modern kangaroo. This prompted Leidy to question Owen's reconstructions. He suggested that *Hadrosaurus* supported itself on its back legs and tail while it was browsing on trees, but may have sunk down on all fours in order to walk around. Leidy's surmise was also based upon some inspired guesswork, since he could not be sure of the length of the tail, or of the proportions of much else of the body. Nevertheless time has proved him to have been far more correct than Owen with respect to this type of dinosaur.

Leidy was soon to be supported by his student Cope, following his discovery and description of *Laelaps* in 1866. This dinosaur had an even greater difference between the lengths of front and hind limbs and confirmed Leidy's tentative suggestions about the kangaroo-like form of these reptiles most firmly.

Anatomists in Britain were not slow to take up the ideas developed in America. One of the leading anatomists, and a keen rival of Owen, was Thomas Huxley. Owen and he had clashed bitterly over Charles Darwin's theory of evolution by natural selection, following the publication of Darwin's *On the Origin of Species* in 1859. Dinosaurs provided Huxley with another opportunity to attack Owen's anti-Darwinian interpretations, and at the same time to work out his own ideas on the subject in a distinctly evolutionary light. By 1868 Huxley was able to draw many lines of evidence together in order to dismantle Owen's elephantine reptiles. Three-toed, bird-like footprints had been discovered in the early 1850s in Sussex in the rock formations in which *Iguanodon* was known to occur; and Owen himself had in 1858

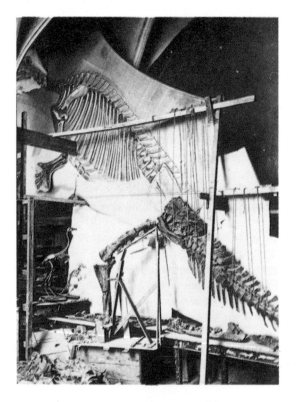

The massive task of sorting and reconstructing the rich finds of Iguanodon *skeletons from Bernissart was undertaken in the large space of the Chapel of St George of Nassau. The reconstructions were carried out with reference to the skeletons of a cassowary and wallaby, shown in this photograph. The resulting skeletons presented* Iguanodon *as a bipedal animal with bird-like and kangaroo-like features.*

described the newly discovered foot of *Iguanodon*, which was three-toed. Leidy and Cope had discovered bipedal (upright walking) dinosaurs in America which were otherwise similar to *Iguanodon* and *Megalosaurus*, while Edward Hitchcock had found large bird-like footprints in the Triassic rocks of Connecticut. And in 1861 the tiny bird-like dinosaur *Compsognathus* had been discovered in Jurassic rocks in Bavaria. Put all these facts together and a rather different picture of the dinosaurs begin to emerge. *Iguanodon* had bird-like feet and left bird-like footprints in the rocks. Dinosaurs very like *Iguanodon* seem to have walked habitually on their hind legs in America. The giant bird-like footprints from the Triassic rocks of the Connecticut Valley were probably left by dinosaurs walking on their hind feet rather than by giant birds (though the smaller prints there might well have been left by true birds, for by now *Archaeopteryx* had been found in Jurassic rock). Finally, not all dinosaurs were large; some, such as *Compsognathus*, were small and delicate, and extremely bird-like in appearance.

The connection seemed obvious to Huxley. Dinosaurs were far from being elephant-like reptiles, they were bird-like. Not only that, but he proposed that the ancestry of birds was probably to be found among the dinosaurian reptiles – the similarities were too many and too close to suggest any other interpretation.

Within the decade Huxley's views about the bipedality and bird-like appearance of *Iguanodon* were entirely vindicated by the discovery of complete skeletons of *Iguanodon* at Bernissart in Belgium. Dollo's work on reconstructing these dinosaurs is interesting in many ways, not least because in the photographs which were taken of the first skeleton of *Iguanodon* you can see skeletons of a wallaby (a small relative of the kangaroo) and a cassowary (a large flightless bird) alongside for reference and guidance. It is plain that Dollo was strongly influenced by Leidy and Huxley; and the evidence for his view was overwhelming. The general proportions of the body of this dinosaur are amazingly similar to those of a kangaroo, confirming Leidy's interpretation. On the other hand, when looked at in detail the feet and legs are extremely bird-like in shape; and – most importantly – the bones of the hips are arranged very much like those of birds (for this is an ornithischian or "bird-hipped" dinosaur), confirming Huxley's views as well.

At about this time dinosaurs began to be discovered in great numbers and variety in Europe and North America. The finds brought a rapidly rising tide of information; but instead of this leading to a progressively better understanding of the nature of dinosaurs and their biology, it seems to have led to an increased diversity of opinion and interpretation, and not a little confusion in the scientific community.

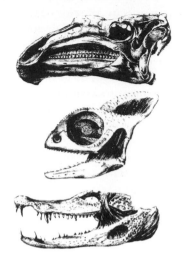

Dollo's studies

TOP: *The bony tendons criss-cross the spines of the back and helped to support the weight of the dinosaur.*
ABOVE: *Dollo used diagrams of the muscles of the legs and hips of birds in order to model the leg muscles of his dinosaurs.*

ABOVE: *Meticulous drawings of the skulls of living and fossil reptiles were used to build up a picture of feeding in dinosaurs. Here he compares (from the top)* Iguanodon *a chameleon and an alligator.*

Dollo and the biology of dinosaurs

Louis Dollo's intention during his lengthy studies of the *Iguanodon* skeletons of Bernissart was to understand the biology and way of life of these creatures as completely as possible.

He was, for instance, extremely interested in the shape and arrangement of the hip bones and legs. He had seen that the tall spines of the backbone were laced with a trelliswork of long bony strands (what we today call "ossified tendons"). He dissected flightless birds and looked in detail at the structure of the backbone, and its supporting ligaments and tendons; he even referred to some of Owen's very fine early descriptions of the anatomy of the kiwi. Dollo also carried out meticulous studies of the muscles of the legs of birds. As a result of this work he developed the idea that he could interpret the shape of the leg bones of *Iguanodon*, and especially the processes – ridges and bumps – for the attachment of the larger muscles. He reconstructed some of the leg muscles of this dinosaur, and suggested that the bony tendons along the sides of the spines in the tail were designed to withstand the enormous pull generated by the leg retractor muscles (those that moved the legs).

Dollo was trying to understand why this animal was the shape it was and how it might have moved on land. He did the same with the skull, looking again at living reptiles (lizards and crocodiles) and carefully noting the arrangement of their jaw muscles, then using this to try and understand the pattern of muscles which operated the jaws in this dinosaur. Not only was Dollo keen to understand the way in which the body of this dinosaur worked, he also wished to place it in its correct ecological context. He regarded *Iguanodon* as a high arboreal (tree) browser, using its height to full advantage to reach the upper foliage. In 1923, in the last scientific paper that he published on the subject of *Iguanodon*, Dollo attempted to show that the living ecological equivalent of this dinosaur was the giraffe.

This clever and thoughtful work was the foundation of much work we today refer to as paleobiology (which means literally "the study of the life of ancient organisms" – with the stress most definitely on the word "life"). This is a branch of paleontology devoted to developing an understanding of the conditions under which organisms lived in the past, and the way in which fossil animals may have moved, eaten, and behaved when alive.

Stupid, slow-moving reptiles?

The image of dinosaurs which began to appear through the work of Cope, Huxley, and Dollo was rather a lively one. Their dinosaurs were pictured in active postures, and the association which was made with birds – a particularly lively group of animals – did nothing to detract from this general image. So it is curious to find that during these very same years other paleontologists, using different dinosaurs, put forward a completely different view of dinosaurs as slow, lumbering and stupid creatures. These views, which harked back in some respects to those of Richard Owen, gained very wide acceptance – and indeed still color the views of many today.

The "low" theory seems to date back to the great dinosaur discoveries in the American Midwest in the late 1870s and early 1880s. Among the many imperfect remains that were hastily recovered from the Wyoming and Colorado quarries at the height of the Cope–Marsh rivalry, a few at least were moderately complete. Among the most dramatic examples were the giant sauropods. Large numbers of bones of a creature which Marsh called *Morosaurus* ("stupid lizard"), and Cope at the same time was calling *Camarasaurus* ("chambered lizard"), were discovered and described. Marsh was particularly struck by the ridiculously small size of the head of such a huge animal. In 1883 enough of another sauropod which Marsh called *Brontosaurus* ("thunder lizard" – later renamed *Apatosaurus*, "headless lizard") was known to allow him to make an early reconstruction of a sauropod dinosaur,

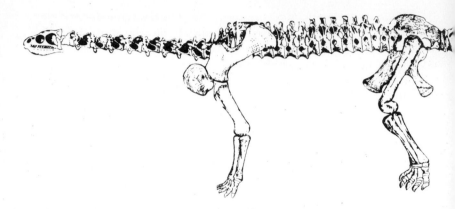

although the remains lacked a head. His drawing of the dinosaur showed that it was colossal by any standards. In Marsh's own words:

> The diminutive head will first attract attention, as it is smaller in proportion to the body than in any vertebrate hitherto known. The entire skull is less in diameter or actual weight than the fourth of fifth cervical [neck] vertebra. A careful estimate of the size of *Brontosaurus,* as here restored, showed that when living the animal must have weighed more than twenty tons. The very small head and brain, and slender neural cord, indicate a stupid, slow moving reptile.... In habits, *Brontosaurus* was more or less amphibious, and its food was probably aquatic plants or other succulent vegetation. The remains are usually found in localities where the animals had evidently become mired.

With these few words Marsh set an image for dinosaurs which lasted for almost a century. Cope, of course, had a different opinion, suggesting that these animals were land-living and rather giraffe-like, using their long necks to reach high into the trees; however, even he was soon convinced by Marsh's persuasive view, and accepted the semi-aquatic marsh-dwelling model for these giant dinosaurs. After all, they appeared to have been far heavier than any of today's land animals. Combine this with the apparent weakness of the wrist and ankle joints in particular – not all the bones are present or well formed – and it would appear that animals such as this would have had to move about in water. The water would have buoyed up their bodies, reducing the weight carried by the limb joints. It seemed likely that any attempt to walk on land would seem to have made their joints grind excruciatingly. What was more, these dinosaurs seemed to have the bony openings of their nostrils perched upon the top of their heads. Living creatures which have their nostrils similarly placed include whales, sea cows, hippopotamuses and crocodiles, all of which are aquatic or semi-

E. D. Cope directed the first life-sized drawing of a dinosaur based on his discoveries in Colorado. This stupendous illustration was displayed at a meeting of the American Philosophical Society in 1877.

aquatic – the high nostrils allow these animals to breathe while submerged with only the tops of their heads exposed.

From this viewpoint the incredibly long necks of these creatures can be seen to have had an altogether different function: they acted as snorkels so that the animals could breathe while wading in deep water. In addition, the amphibious habit meant that these creatures could wallow near the margins of lakes and streams and feed on abundant water plants along the shore or in the shallows. This peaceful and slow existence seems to accord with the diminutive brain of these gigantic animals.

Some of Marsh's other discoveries seemed to confirm this view. *Stegosaurus*, though considerably smaller than his *Brontosaurus*, was also a rather clumsy-looking animal, and was again remarkable for the small size of its head and brain. Indeed, Marsh noted that the brain was considerably smaller than a proportion of the nerve cord in the spine above the hips. The second "brain" was seen as a relay station for messages from the tail and a backup to the true brain in the head. That conjured up the image of an animal that might be bitten on the tail and not even realize it – a classic example of the stupid dinosaur.

Contorted monsters

The giant sauropods became the centre of an even more curious disagreement in the early years of the twentieth century. Marsh's reconstruction of *Brontosaurus* seems not to have provoked much criticism or discussion, which is a little perplexing when you bear in mind the previous furore over the posture of Owen's dinosaurs. The actual shape of the main body of *Brontosaurus*, if you ignore the extreme length of the neck and tail, is not too different from the posture of Owen's Crystal Palace models – and yet Cope, Leidy and particularly Huxley had poured scorn on Owen's elephantine reptiles. But there was no getting around the anatomy of the sauropods. With *Iguanodon*

These huge sauropod dinosaurs were once thought to have lived in deep water, using their long necks like snorkels.

and *Megalosaurus*, it had been a relatively simple matter to provide a bipedal posture. But no matter how you look at sauropods, it is difficult to imagine them striding about on their hind legs alone. To some extent, therefore, Owen had been vindicated.

Sauropods such as Marsh's gained worldwide recognition through the sponsorship of Andrew Carnegie, who had casts of *Diplodocus carnegie* made and sent to various parts of the world. Close examination of these specimens by interested scientists and others produced a welter of opinions about the correctness, or otherwise of the mounted skeleton.

Criticism of the dinosaur reconstruction focused on the positioning of the limbs. It was pointed out in early reviews that since *Diplodocus* was a gigantic lizard, surely it should have its legs mounted in the usual posture for a lizard – sprawled out from the sides of the body. In this way the great weight of the body and the long neck and tail could rest with much greater ease on the ground, rather than being held aloft in such an "unnatural" position. In 1906 a group of assistants at the American Museum of Natural History made a model of this dinosaur in just such a position, but when it was exhibited it was roundly condemned as impossible by a range of paleontologists. However, the argument did not die at this point.

The paleontologist Oliver Hay quite independently argued that *Diplodocus* could not possibly have walked with upright legs. His reasoning was deceptively simple – and ignored all the basic principles established by Cuvier and Owen years before. He knew that these creatures were reptiles rather than mammals, so it was preposterous to try and reconstruct them after the fashion of elephant-like lizards. First, reptiles (or rather, today's reptiles) do not walk with their legs directly under their bodies. Second, they were so large and heavy that their immense weight would have forced them to flop down on their bellies. Third, if these animals did, as Marsh had argued, inhabit marshy country, they may have better been able to move around by slithering on the ground; had they tried to walk on such soft ground their feet would instantly have sunk in and they would have perished miserably. In 1910 Hay backed up his argument with his own restoration of *Diplodocus* crawling in their marshy habitat. Hay was not alone in his criticism, being supported by several paleontologists in Germany, among them Gustuv Tornier, who produced his own reconstruction of *Diplodocus* in a lizard-like posture.

In reply Dr Holland, Director of the Carnegie Museum and the man responsible for supervising the restoration of *Diplodocus*, poured scorn on these observations and the contorted monstrosities that had been created by Hay and Tornier. He pointed out, quite correctly, that if this dinosaur was reconstructed with its legs sprawling out sideways as had been suggested, then it would have needed a deep rut in the ground to accommodate its very deep rib cage. Additionally, to make it assume such a posture, all the joints of the legs would need to be dislocated.

Close studies of the joints of the shoulders, hips, legs, and feet proved to Holland's satisfaction that the legs of this animal were held upright. The argument faded after this, and no skeleton was ever dismantled and rearranged in a lizard-like pose; but final proof of Holland's argument did not appear until the late 1930s. At that time Roland T. Bird discovered a trail of footprints left by a large sauropod dinosaur at Glen Rose in Texas. The stride of the beast was a good 12 ft (3.6 m), yet the width of the track, the distance between the left and right feet, was a mere 6 ft (1.8 m). If the sauropod had been crawling along with sprawled legs the width would have been much greater. Here at last was proof that sauropods walked upright.

In 1906 Otto and Charles Falkenback of the American Museum in New York created a model of a "creeping" brontosaur. When exhibited it appears to have been roundly condemned, but the arguments about the creature's posture continued.

99

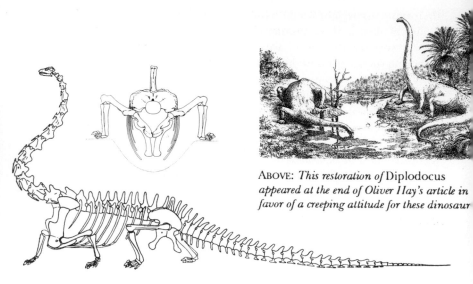

ABOVE: *This restoration of* Diplodocus *appeared at the end of Oliver Hay's article in favor of a creeping attitude for these dinosaur*

CREEPING SAUROPODS - OR CONTORTED MONSTERS?
Holland showed that restorations of creeping sauropods such as Tornier's (above) and Hay's (right) were absurd, requiring a trench to accommodate the deep chest (top).

Slothful meat-eaters?

The view of Marsh that dinosaurs were slow-moving and slow-witted creatures was one that struck a deep chord with many vertebrate pale-ontologists. Experience with living reptiles had shown them to be rel-atively inactive creatures, spending long periods of time resting, and unable to sustain high levels of activity such as running for more than a few minutes.

We have seen that bipedal herbivores (*Hadrosaurus, Iguanodon*) and carnivores (*Megalosaurus, Laelaps*) were represented in an active, even leaping, pose in the 1870s. Yet by the second decade of the twen-tieth century that picture was to become less clear. Lawrence Lambe, who assisted with the Red Deer River expeditions of the Canadians, was rewarded in 1913 by the discovery of the skeleton of a Late Cretaceous carnivorous dinosaur which he named *Gorgosaurus* (it was later called *Albertosaurus*). Here was a giant but, if looked at carefully, rather graceful dinosaur 29 ft (8.8 m) long, standing about 11 ft (3.4 m) tall and weighing perhaps 1 1/2 or 2 tons.

What sort of lifestyle would this dinosaur have had? Lambe's inter-pretation conformed exactly to the slothful, stupid image created by Marsh and later workers. It was that *Gorgosaurus* would have spent much of its life lying prostrate on the ground. It could walk erect on its hind legs – that was clearly shown by the shape of the bones. However, to be provoked into such energetic activity the gorgosaur would have had to be very hungry. To support this view Lambe was even able to show the bones of the pelvis were so constructed that it was able to support itself when lying on the ground, without its belly and lungs being crushed by its immense weight. As for food, obviously this was dinosaur meat but, rather than being an active hunter of food, *Gorgosaurus* was in Lambe's view a scavenger. It fed, he claimed, on "carcasses found or stumbled across during its hunger-impelled wanderings".

In contrast to Lambe's interpretation, another and far better known large, carnivorous dinosaur was being discovered in Late Cretaceous rocks and described in America. One of the many expeditions sent out by the American Museum of Natural History in New York during the early years of the twentieth century found some large skeletons in northern Montana. By 1905 enough material had been collected and prepared back in New York to allow a description to be made by the museum's director, the paleontologist Henry Fairfield Osborn. It was christened with the royal name *Tyrannosaurus rex* in recognition of the fact that it was the largest land-living carnivore then known to have walked the earth. With a huge skull nearly 4 ft (1.2 m) long bearing enormous 8 in (20 cm) long teeth, and powerful hind legs ending in massively clawed feet, this animal was seen as the ultimate killing machine. Indeed it seemed to have sacrificed its arms in its quest for perfection as a killer; these were so reduced that they could not even reach its mouth.

The image created by the work of Lawrence Lambe was of a huge scavenging dinosaur such as Albertosaur *which spent most of its life prostrate on the ground, only lurching to its feet when hunger became too great.*

Tyrannosaurus has always had a much more active image, as shown in Charles Knight's classic painting, where it grapples with a live Triceratops.

Osborn and his colleagues concluded that this animal was the implacable foe of creatures such as *Triceratops*, the giant frilled and horned ceratopian dinosaur, and of the hadrosaurs, whose remains had been found in similarly aged rocks. As to the question of exactly how active these dinosaurs were, Osborn and his colleagues were clearly in a quandary. Barnum Brown, who had found and collected the specimens, thought that *Tyrannosaurus* was "active and swift of movement when the occasion arose" and when he later discovered the similarly built *Gorgosaurus* from the Red Deer River he restored it in a running pose, chasing a herd of duck-billed hadrosaurs. Osborn and Matthew were, however, a little more restrained in their view, suggesting that *Tyranosaurus* would have fed in the manner of a lizard. It would have made a lunge, snapping with its teeth and gouging with its claws until the prey was overcome. If combat was involved between predator and prey, Matthew suggested, it would have been a slow-motion affair, however bloody the outcome.

Naturally not all paleontologists had the same views about the life and nature of dinosaurs at any one time. There is almost always a diversity of opinion among active and intelligent workers. However, they tend to work within what we might call a consensus view; and in the early decades of this century the consensus was that the dinosaurs were large, slow-moving, and slow-witted creatures. Yet this image did not fit comfortably with a number of earlier observations. Not all · dinosaurs were big: Huxley had used the example of the tiny *Compsognathus* in the 1870s, and had also described a tiny British dinosaur, *Hypsilophodon*. Nor did they all seem either slow-moving or slow-witted, particularly the small, nimble-looking ones.

DINOSAURS IN ACTION

In the past few years ideas have turned around again. Scientists are now convinced that dinosaurs were neither slow nor noticeably stupid. Their work tends to fall into two spheres of research: one is concerned with insights into the life and activities of dinosaurs, as revealed by their structure; the other follows more theoretical lines of argument.

How did hadrosaurs live?

If anyone had doubted that dinosaurs were reptiles, that fact was confirmed in dramatic style by Charles Hazelius Sternberg and his sons, who discovered a skeleton of the hadrosaur *Anatosaurus* in Kansas, around which was preserved the impression of its skin. This "mummified" dinosaur had clearly perished in a sandstorm; buried in dry sand, its body had slowly dried out undisturbed by scavengers, leaving its skin to dry to a hard parchment. The fine sand became molded against the skin, retaining a permanent record of its shape in the rock.

The skin appeared to be fairly typically reptilian in appearance, with small, rounded scales separated from each other by areas of thinner, more flexible, skin. This neatly confirmed the conclusions of paleontologists whose anatomical studies had revealed so many other reptilian characteristics.

Hadrosaur skin was not quite the same as that of modern lizards and snakes with their dense overlapping scales but was more like a mosaic of scales of varying sizes. The scales also seem to vary in size over the body: in some places, such as across the back where it was exposed to the sun, the scales were quite large and prominent, and in other, less exposed places they were small. The skin impression was also preserved around the bones of the front paw or hand, and seemed to show that there was a paddle-like mitten of skin covering the hand. This fitted well with the ideas about the habits of these creatures. It had been supposed that they lived in or around water. The presence of a broad, duck-like beak which gave them their popular name "duck-billed dinosaurs" suggested that they may have dabbled in water, collecting aquatic weeds. The webbed hands suitable for swimming seemed to support this idea even more strongly.

Shortly after this discovery even more startling discoveries of hadrosaurs were made, this time in Canada along the Red Deer River. In 1913 it was again the indefatigable Charles H. Sternberg and his sons who found the first of a new type. This was given to Lawrence Lambe at the Canadian Geological Survey, who described it as *Stephanosaurus* ("crowned lizard") because of the tall crest which rose from the crown of its head. Later this dinosaur was renamed *Lambeosaurus* in honor of Lambe's work. Very soon a variety of crested hadrosaurs began to be found: some had relatively flat heads, with perhaps a thickened nose area as in *Kritosaurus*; some had broad, rounded helmet-like crests, for example *Lambeosaurus* and *Corythosaurus*; others had spiky crests, like *Saurolophus*; and yet others were ornamented with long, tubular swept-back crests such as that of *Parasaurolophus*. Fascinating though these new kinds of hadrosaurs were to describe at first, the question remained of what the various crests were used for. Obviously they were very convenient because they allowed paleontologists to tell hadrosaurs apart – but that surely was not their function in life.... or was it?

In 1929 the eccentric paleontologist Franz Baron Nopcsa (pronounced "Nopsha") became intrigued by hadrosaur crests and their function. It had already been suggested that these crests were simply horns, much like the horns and other adornments found on the heads of cattle, sheep and goats. What Nopcsa did was to take this line of reasoning one step further; he suggested that the crests were a sign of the sex of the creatures. It is a well known biological fact, established by Charles Darwin, that male species are often more elaborately decorated than the female – this is particularly well known in birds, where the male is often flamboyantly colored, while the female is dull and unobtrusive. Nopcsa suggested that the crests of hadrosaurs have a parallel in the plume of feathers on the bird's head. Thus those hadrosaurs with large crests, such as *Parasaurolophus* and *Corythosaurus* would be the "males" of the species whose "females" were *Kritosaurus* and *Anatosaurus* respectively.

Original though this suggestion was, it did not gain very wide acceptance. It did not take long to realize that the crests were not sim-

HADROSAUR MUMMY

TOP: *Dinosaur mummies are extremely rare. This one collected by Charles Sternberg is of the hadrosaur* Anatosaurus *which probably perished during a sandstorm. The windblown sand packed around the cadaver preserved a marvellous impression of its skin*

BOTTOM: *The skin of hadrosaurs varied depending upon the area of the body which it covered in life. The scales seem to have formed a pattern like a mosaic of tiles, rather than an overlapping pattern as they do in many living reptiles.*

ply adornments to the head but had complex cavities inside; and, to make matters worse, it became obvious that some of Nopcsa's "male" hadrosaurs were only found in one part of the world, while his "females" were found elsewhere.

PARASAUROLOPHUS WALKERI
The crests of some hadrosaurs are really quite spectacular structures. The long, curved, tubular crest of this dinosaur may, or may not, have had a sail-like flap of skin running down onto its neck, but the head and crest were very probably brightly colored and distinctive.

Since hadrosaurs evidently lived in or around water for much of their lives, there were several ideas connected with the notion that these crests were some form of breathing tube. In 1938 Marton Wilfarth suggested that many dinosaurs had lived at a time of universal flooding, and proceeded to identify numerous features of their anatomy which pointed to adaptations for living in moderate or deep water. In the case of hadrosaurs, the hollow crest and the duck bill were seen as areas for the attachment of an elephant-like trunk. This arrangement allowed the animals to breathe through what would have been a muscular snorkel, which could also be used like an elephant's trunk for gathering up bottom weed. Again, fascinating though this idea was, it did not attract many converts. The evidence for muscle attachments for the supposed trunk could not be found, and it seemed senseless to evolve a trunk for gathering food when these animals had perfectly serviceable beaks. The theory also failed to explain why the passages inside the bony crest were so complicated.

Variants on this idea associated with the idea of linking the crest to a watery lifestyle did, however, gain some measure of acceptance. For example, the snorkel idea was adapted rather cleverly. In the case of the tubular-crested hadrosaur *Parasaurolophus,* it seemed "obvious" that when the beak was dipped into water, the long, tubular crest would have projected into the air jest like a real snorkel. While this did not apply to all hadrosaurs, because not all had a long tubular crest, it did seem like a good idea for at least this kind. Unfortunately the theory had one fatal flaw. The top end of the crest needed to be open if the snorkel theory was to work; but, sadly the crest was closed off with bone and was never open in life.

Another novel idea come from Sternberg himself. Many of the crests seemed to have an S-shaped loop inside; this, he suggested, acted like an airlock to prevent water from rushing into the lungs when their head were submerged for feeding on water weeds. There are two problems with this theory. First, the pressure of water on the airlock when the head was submerged would have been far too great for such a simple device. Second, all living reptiles and mammals that submerge their heads are capable of closing off their nostrils with flaps of skin which act as valves – and it would be very surprising if hadrosaurs had not done the same, thus making an airlock unnecessary.

The third variation on the theme came about in the 1940s with the idea that the cavities of the crests acted like the air tanks of a scuba diver. The theory was that the hadrosaur took a large gulp of air before submerging, and stored some of the air in its crest for use later in the dive. Again the idea is appealing at first sight, but further thought suggests it to be fraught with problems. First, the volume of the crest is tiny compared to the likely size of the lungs in these creatures, rendering the reservoir of very little value. Second, it is very difficult to imagine how the air might have travelled from the crest to the lungs without the presence of a special pumping mechanism, and the consequent risk of air pressure changes causing the head of the hadrosaur to collapse!

Eggs and egg thieves

During the 1920s another dramatic discovery was made, this time in Mongolia, by a team of explorers from the American Museum of Natural History. What they discovered, largely by accident because they had not gone to the area specifically to look for dinosaurs, were the nests and eggs of dinosaurs. The oval, relatively long eggs were found in neat circular nest-like arrays of twenty or more at localities near the beautiful red cliffs, known as the Flaming Cliffs, at Bain Dzak in southwest Mongolia. Associated with these nests were a large number of dinosaur skeletons of a new type of small ceratopian dinosaur, which was named *Protoceratops* ("early horned face").

These discoveries revealed a number of things. First, and most important, they provided the first-ever proof that dinosaurs laid eggs.

Fight to the death

RIGHT: *This
remarkable discovery
made at Bain Dzak
in Mongolia shows
two dinosaurs
fossilized in mortal
combat. The
predatory dinosaur*
Velociraptor *(on
the right) can be
seen grasping the
head of its*

unfortunate victim Protoceratops. *Notice how the legs of the predator are drawn
up under the body. A close relative of* Deinonychus, Velociraptor *had large,
sharply hooked claws on its hind feet which it used to disembowel its victim with
vicious kicks to the unprotected belly. This is one of the only examples known in the
fossil record of dinosaurs caught in combat.*

BELOW: Psittacosaurus *can be seen in the left background, facing a feeding*
Velociraptor. *In the foreground a* Protoceratops *by its nest has caught the foot
of a fleeing* Oviraptor.

The discovery created a minor sensation at the time, bringing enormous publicity to the expedition members and the museum. This was fortunate because the expedition, at least in its early years, had largely failed to find what it had set out to discover in the first place, which was the remains of early humans! The evidence, like the early discovery of the scaly skin of hadrosaurs, confirmed the reptilian nature of dinosaurs. However, in addition to the eggs a large quantity of *Protoceratops* bones and skeletons were discovered, which revealed other aspects of their biology. Individuals of various sizes and ages were discovered, which allowed growth to be studied for the first time in any dinosaur. It also became clear that there were differences between the skulls of some individuals which could only be satisfactorily be explained by their being of opposite sexes – again a first in dinosaur studies.

Other dinosaurs living alongside *Protoceratops* were also identified, including a lighter, bipedal ceratopian dinosaur, which lacked evidence of horns or a frill and was named *Psittacosaurus* ("parrot lizard"). And living amongst these herbivores there were predators. *Velociraptor* ("speedy predator") was a nimble predator with long legs, long grasping arms, and sharply toothed jaws. More curious still among the predators were the rather puzzling remains of another animal. This was another lightly built predator which was smaller than *Velociraptor* but, unlike most predatory types, had a curiously snub-nosed skull with toothless jaws – in fact it appeared to have a beak more like that of a turtle than of a typical predatory dinosaur. The habits of this dinosaur would have presented a problem had it not been for an extraordinary discovery. The flattened skull of one of them was discovered on top of a nest of *Protoceratops* eggs. It seemed to have been caught in the act of stealing them – and so it was named *Oviraptor*, which means "egg predator." It can only be guessed at, but the circumstances point to an enraged *Protoceratops* returning to its nest and stamping on the would-be egg thief. Certainly a diet of eggs would suit a toothless predator.

Clues from footprints

In 1938 Roland T. Bird, a dinosaur collector working for the American Museum, discovered dinosaur tracks at Glen Rose in Texas. The prints were not only those of the three-toed variety, which had been known for over a century (see page 67), but also the first recorded footprints of a large sauropod dinosaur, which were over 1 yard (1 m) across. During the following year a team of collectors was sent out to Glen Rose to dig up a complete trackway from this area. They made an unexpectedly important discovery. Alongside a particularly fine sequence of prints left by a sauropod were the unmistakable bird-like prints left by a large predator such as *Allosaurus*. This seemed to provide dramatic proof of the conflict between predator and prey. The trackway was unfortunately incomplete. Was it just coincidental that the tracks seem to follow one another? Was the allosaur tracking the sauropod? Though it can never be proved, it is promising circumstantial evidence for a predator stalking its prey. That hardly fits with the image which Lambe conjured up of the slothful predator.

Bird's discovery established two things about dinosaurs. First, as we have already seen, sauropods made narrow tracks, with their feet close together and directly beneath the body, just as Owen had predicted, rather than having legs sprawled out sideways like those of lizards, as had been suggested in the early decades of the twentieth century (see page 67). Second, large dinosaurs may have hunted their prey actively, rather than waiting for them to drop down dead so that they could feed on the rotting carcass.

A few years later, in 1944, Bird added further confusion to the debate over the habitats in which the giant sauropods preferred to live with the discovery of an even more remarkable series of tracks in Bandera County, Texas. They were on what might have been the

RIGHT: *The footprints at Glen Rose discovered by R. T. Bird show the huge prints of sauropods, with smaller, bird-like prints of predators.*

The Glen Rose footprints suggest a scene with an allosaur stalking a much large brontosaur along a mud flat in the Late Jurassic.

Some footprints show that sauropods did take to the water, floating at the surface and poling themselves along using their front feet.

muddy bottom of a large lake, and consisted of a row of front footprints of a large sauropod, similar to *Diplodocus*. No hind footprints were found at first, and it seemed that the dinosaur must have been floating in deep water, and kicking itself along the bottom using front feet alone. A little further along the track sequence there is a single hind footprint, after which the front feet show a sharp change in direction. The hind feet were clearly held clear of the bottom as the animal floated along and were used occasionally for steering! Here was clear evidence that sauropods could and did swim.

So where did sauropods prefer to live? The consensus at this time would have favored a life in water, with very rare excursions onto land. This was supported by Bird's "floating" trackway; by the enormous size of the creatures, which suggested that they needed to live in water in order not to collapse under their enormous weight; and belief that they would have lived on succulent aquatic plants; and their heads with nostrils mounted on the top.

Are reptiles "cold-blooded"?

One feature of lizards that had become fixed in the minds of many biologists around the turn of the century was that they were "cold-blooded" – or, to be more precise, that they had no internal control of their body temperature. This is important because an animal's muscles, nerves, digestion, and indeed its whole metabolism, need to be reasonably warm in order to work at full speed. A cold animal was sluggish. Oddly, a "cold-blooded" animal is also in danger of overheating if it stays too long in the sun, and can die as a result. "Warm-blooded" creatures can cool themselves by panting and sweating.

Experiments with lizards in glass laboratory tanks seemed to suggest that their body temperature varied precisely with the tempera-

Modern crocodiles are able to keep a relatively stable body temperature simply by being big - an advantage that was not lost in the dinosaurs.

ture of the air around them. The warmer the air, the warmer their bodies and the more active they could be; and conversely in cooler air they would be less active. These observations were widely accepted, and supported the notion that these types of creature were less effective – indeed some would say more primitive – than "warm-blooded" mammals and birds, which are able to keep their bodies warm and remain active irrespective of their air temperature. Dinosaurs, being reptiles, were imagined to be similarly "cold-blooded" and equally primitive.

This distinction between "cold-blooded" and "warm-blooded" creatures turns out to be far less simple than was first thought. Measuring the temperatures of lizards in glass tanks in laboratories turned out to be misleading. Experiments in the wild produced completely different results. Lizards actually maintain a surprisingly high body temperature in their normal environment. The majority remain between 95 and 120°F (35 and 42°C) during daylight hours, which compares well enough with humans at 98.6°F (37°C). They are to all intents and purposes warm-blooded!

How do they do this? Mammals and birds produce body heat internally through biochemical reactions – and are for this reason more correctly called endotherms ("inside heat"). Reptiles, by contrast, rely on external sources of heat to keep their bodies warm and are more correctly called ectotherms ("outside heat"). Lizards, for example, bask in the sun, stretching out so as to absorb as much heat as possible and transferring the heat around the body in the blood, which acts like the hot water in a central heating system. During the day a lizard will move between hot, exposed places and shady areas to warm up or cool off as required, and in this way will keep its internal temperature reasonably constant. Obviously there is a drawback to being an ectotherm: on a cool cloudy day it can be difficult to warm up, and at night (except in the tropics) it can become too cool for the animal to remain active. That explains why reptiles are much more widespread in the tropical and subtropical regions of the Earth today, where warmth of the sun or air is much more dependable than in the cooler regions at higher latitudes.

You may wonder whether this is really relevant to the way of life of dinosaurs. The answer is that it is extremely important, as was pointed out in 1946 by a group of scientists: Professor Edwin Colbert, a dinosaur paleontologist of world renown; Charles Bogert, a reptile paleontologist (both of these worked at the American Museum), and Raymond Cowles in the University of California. These men did some experiments to compare the body temperatures of large and small living reptiles during a day and night cycle. They found, perhaps not surprisingly, that the body temperature of small alligators tends to change much more rapidly, both during the daytime, and between day and night, than does the body temperature of large alligators. For example a small alligator of about 1 ft(30 cm) in length tethered in the sun increased its body temperature five times more quickly than a 4 ft (1.3 m) long creature. The reason for this is that the volume of a large creature is far greater in proportion to the surface area of its body than in a small animal. The amount of heat a body holds depends on its volume, and the speed at which heat can pass in or out depends on its surface area. So a large body heats up more slowly than a small one in the same conditions. The rate of temperature change in the bodies of alligators does not necessarily compare well with that in dinosaur bodies, but it does provide a very interesting pointer.

Simple observations on relatively small living reptiles can be scaled up to apply to a dinosaur weighing as much as 10 tons. Colbert and his colleagues suggested that such large creatures could have taken something close to 48 hours to raise their temperature by a mere 1°F (or 86 hours for 1°C). Put another way, if a dinosaur's body cooled, for whatever reason, by just one degree it would need to bask continuously in the sun for all that time in order to bring its temperature back up to normal! This is clearly absurd, and is not really the point at all. The crucial point, and the one that Colbert, Cowles, and Bogert made, was that it can equally be argued that huge dinosaurs, because they have an immense bulk and a relatively small skin surface compared to that bulk, would have taken an enormously long period of time to cool down by one degree! Given that dinosaurs did live at a time of very mild, generally warm climatic conditions (see page 40), their huge size might have meant that they could have easily retained their body heat overnight.

If Colbert, Cowles, and Bogert were correct they could really provide interesting possibilities for how active large dinosaurs could be.

Dinosaurs could now be seen as thermally stable reptiles capable of keeping their bodies warm not by their internal biochemistry – as in the case of mammals and birds – but simply by being very large.

Startling suggestions such as these provoked a great deal of controversy among scientists, and within a year Colbert and colleagues had trimmed their estimate of several days for a large dinosaur to cool by one degree, down to "several hours." Indeed the impact of these initial observations was greatly diluted by the fact that their estimates became progressively shorter and shorter over the years to the point where one of the authors suggested that a 10-ton dinosaur would have been capable of raising its body temperature by 3.5°F (2°C) if left to bask in the hot sun for little more than an hour.

The end of the snorkel

In the early 1950s Kenneth Kermack of University College, London, who was to become a noted authority on early mammals, produced an argument which directly contradicted the idea that sauropods lived in deep water and used their necks and heads as snorkels. Kermack studied the results of some research on human breathing underwater and applied them to sauropod dinosaurs. Unlikely though this may seem, it was important. In 1951 Kermack produced an article in which he argued that it would have been physically impossible for sauropods to have breathed air into their lungs with their bodies submerged at even a relatively modest depth. Human experiments had found it virtually impossible to breath air through a snorkel at depths greater than 3 ft (0.9 m), and it was only possible to breathe comfortably at a depth of 10 in (25 cm)! The reason for this was the pressure of water around the chest, which at even a small depth is much higher than that of the air at the surface, and literally squeezes air out of the lungs, making it impossible to breathe in. In the case of a long-necked sauropod, whose lungs might be 20 or 30 ft (6-9 m) beneath the water surface, the water pressure on the chest would have been colossal – crushing the lungs and windpipe flat. For a sauropod to breathe at all at such depths the lungs and windpipe would have had to be encased in a cylinder of solid bone, and the muscles which pumped the lungs would have been needed to develop a force amounting to several tons to drag air down into the lungs against the water pressure – so high that it would have ruptured all the linings of the lungs.

The argument does seem very convincing, because the physics of water pressure are absolutely unarguable. Casting around for modern comparisons, whales which can dive to great depths only breathe when their bodies – and lungs – are right up at the sea surface where the pressure is least. Whales also breathe out before they dive so that their lungs are empty.

However, dinosaur experts of the time seem not to have been prepared to accept Kermack's arguments. Professor Colbert, who had been at the forefront of the experiments on alligators and their implications concerning body temperatures of dinosaurs, was quite convinced that Kermack was wrong. He suggested that using a human example of the difficulties of breathing underwater for comparison with dinosaurs was unwise, since their skeletons and physiology would undoubtedly have been totally different. He went on to plead for a special case for dinosaurs, which may have had some "modification" which would have allowed them to breathe while submerged. Such special pleading may now seem extremely suspect, but it clearly did not seem to be regarded in that way at the time. Evidently the consensus view ruled the minds of the dinosaur experts. Most believed that sauropods lived in deep water; the evidence was there in trackways, and in their anatomy, so Kermack had to be wrong!

Dinosaurs and birds

Last, but by no means least, among this roster of people and their ideas about dinosaurs is the Danish paleontologist Gerhard Heilmann. In 1926 he published a long and detailed book, *The Origin of Birds*, in which he examined the problem of the origin of birds from reptiles. While a kinship between reptiles in general and birds had never been seriously doubted – one look at the scaly feet and claws of a bird and the fact that both groups lay eggs with shells seems enough to indicate at least some remote connection between the two groups – there was much disagreement over precisely which group of reptiles were the nearest relatives of birds. Thomas Huxley and others had already discussed in the 1870s a possible dinosaur connection in the origin of birds. As Huxley had clearly shown, some dinosaurs had very bird-like traits, and not all were large, heavy animals; some (and

This small, light, bird-like dinosaur, Compsognathus, *provided compelling evidence of a close relationship between at least some dinosaurs and birds.*

Struthiomimus was the ultimate in bird-like dinosaurs. These extraordinary creatures, which lived during the Late Cretaceous, were not only the size of ostriches, they also looked like them. Their legs were very long and spindly, they had long, bird-like necks, large eyes and a toothless beak - all they seemed to lack were feathers. This illustration is typical of Heilmann's accomplished artworks which lent convincing support to his theory that the lively, fast-moving, bird-like theropods were the obvious ancestors of birds.

he had the marvellous example of the tiny *Compsognathus* to fall back on) were light and delicately built. This could mean one of two things: either, as Huxley suggested, that some dinosaurs belonging to a group similar to *Compsognathus* were distant relatives of birds; or alternatively – and there were many who held such a view – that dinosaurs and early birds had a similar way of life. This latter view could account for a coincidental similarity of appearance. It could be argued by evolutionary biologists that animals living in similar ways tend to develop very similar anatomical characteristics – a process known as convergent evolution. There are examples of modern animals which show convergent features to prove that this is a very real possibility. One of the most celebrated examples is that of the ordinary wolf of Europe, Asia and North America and, on the other hand, the thylacine or marsupial wolf of Australia. To the untrained eye these animals are very similar in appearance, as well as their way of life. They could be thought to be close relatives; but nothing could be further from the truth. The ordinary wolf is actually more closely related to us humans than it is to the thylacine! Both types of creature have, however come to adopt a similar way of life – as fast running, long-legged carnivores – and their courses of evolution have converged so that they now resemble each other remarkably closely.

In the case of bird origins, many thought that convergence was a major factor. As a result the argument moved largely away from the views of Huxley and more towards the idea that the similarities between dinosaurs and birds were a result of evolutionary convergence on a particular style of life – in this case as small, agile, insect-eating creatures. However, what was really needed was a comprehensive and careful look at dinosaur and bird anatomy in order to see if it was possible to disentangle the two lines of argument once and for all. It was Heilmann's work, culminating in his book, which looked at precisely this problem. His detailed review of dinosaur and bird anatomy brought out a remarkable number of similarities between birds and small, carnivorous theropod dinosaurs which at first reading seem to vindicate Huxley completely: ".... striking points of similarity pertained to nearly all parts of the skeleton. From this it would seem a rather obvious conclusion that it is amongst the Coelurosaurs [small theropod dinosaurs] that we are to look for bird ancestors." In partial support of this Heilmann, who was an artist of considerable skill, provided a number of illustrations of dinosaurs in their habitat. All of these are notable for their liveliness of pose – the animals are shown as erect, nimble and fast-moving, superficially bird-like creatures.

However, in his review and conclusions Heilmann stopped short of drawing his "obvious conclusion" – that theropods are bird ancestors – for a single, and to his mind overwhelming, reason. All living and fossil birds have a wishbone, a V-shaped strap of bone that lies at the front of the chest, embedded deep in the wing muscles and connecting the shoulder joints. This single bone is thought to have originated as the two collar bones, now fused together. At the time when Heilmann wrote his book no evidence had ever been recovered of the collar bones of dinosaurs, so there was a problem: how can you suggest that animals with collar bones (in this case birds) evolved from animals which seem never to have possessed such bones (dinosaurs)? The collar bones of dinosaurs appear to have been lost early in their evolutionary history, so for birds to have evolved from dinosaurs collar bones would need to have reappeared, as if by a reversal of evolution. Since evolution is thought to have been irreversible – a dictum known as Dollo's Law (see page 79) – birds simply could not have evolved from theropod dinosaurs. The authority of this verdict on bird origins seems to have been great, for Heilmann's conclusions dominated the views of paleontologists and evolutionary biologists alike for nearly fifty years.

Charting evolution

Thus all the similarities between theropod dinosaurs and birds were supposed to be convergent features developed because of similarities in their ways of life. As a result the search for the ancestry of birds turned away from dinosaurs specifically and towards earlier groups of reptiles – notably among the crocodiles and other related forms of which there was a considerable variety during the Triassic Period.

In moving away from birds and their ancestry, a process started by O. C. Marsh's description of the giant, stupid, slow-moving dinosaurs of the American Jurassic in the 1880s, and culminating in Heilmann's review, dinosaurs seemed to have "lost" the supposed energy and vitality which a connection to birds provided. At least so it would seem, looking back on general books about dinosaurs written during the middle decades of this century. Despite the clear evidence of slender, agile, and clearly fast-moving dinosaurs in the scientific literature, the view was increasingly widely held that all dinosaurs were lethargic creatures.

This all must seem rather strange, because dinosaurs cannot really be equated with modern reptiles – in terms of either their general way of life or their anatomy. No modern reptiles grow as large as some dinosaurs, none walk or run in the way that dinosaurs were able to, and it is certainly the case that no living reptiles can walk habitually on their hind legs. None show those features of dinosaurs which we today link with animals which move very quickly indeed.

Cuvier and Owen grasped the idea very early that there was something special about dinosaurs; but later workers seem to have become less, rather than more, sure of this fact. It is almost as though the marvelous discoveries made in the latter half of the last century confused, rather than clarified things.. Perhaps that it not surprising. Once dinosaurs began to be found in some numbers they proved to be incredibly varied: at one moment Marsh might be describing some new gigantic sauropod, yet within a few months another dinosaur with the size and build of *Compsognathus* might be discovered, or one such as *Stegosaurus* with extraordinary plates down its back. The sheer diversity of types must have been bewildering.

Making sense of such variety by concentrating on some "key" characteristics by which dinosaurs can be contrasted with all other types of animal proved impossible, as we have seen. However, with the new discoveries, research, and interpretations which have occurred over the last thirty years we have begun to make sense of the confusion.

NEW DINOSAURS, NEW IDEAS

The contradictions inherent in most models of how dinosaurs lived came under sharp scrutiny during the 1960s through the work of a number of people, but perhaps the most important contribution has been made by Professor John Ostrom, working with a small team of research students in Marsh's Peabody Museum at Yale University.

More thoughts about hadrosaurs

Ostrom's early work as a student was on the dinosaurs which Leidy, Cope and Marsh cut their intellectual teeth upon: the hadrosaurs. Ostrom had the advantage of a great wealth of new and better preserved material, and devoted himself to studying the intricate details of just the skulls of these creatures. At the time he started his work it was generally thought that these dinosaurs were marsh or swamp dwellers – ideas that could be traced in a direct line back to Joseph Leidy and his early speculations on *Hadrosaurus*. Cope too had reinforced this idea in 1883 by observing that the beak of *"Diclonius"* (his name for what we now call *Anatotitan*) was broad and feeble, and the teeth behind were also weak, a combination that rendered these animals only capable of scooping up and chewing soft aquatic vegetation. Add to this the deep paddle-like tail, and the suspicion that the creatures may have had webbed fingers, and the case seemed to be reasonably secure.

But, looking single-mindedly at the complexity of hadrosaur skulls, Ostrom came to a number of conclusions which completely contradicted the swamp-bound scenario for hadrosaurs. While some hadrosaurs, and particularly the *Anatotitan* which Cope studied, seem to have a broad and superficially duck-like bill at the front of the mouth, the appearance is deceptive. Rather than forming a broad, flat beak adapted for "dabbling" as ducks do, the bones of the beak are strongly attached to the skull, and have a sharp, notched margin. Comparing this to modern animals, the nearest similarity is to tortoises and turtles rather than dabbling birds; turtles have an extremely strong, self-sharpening beak useful for cutting plants and flesh alike.

Further back in the jaw the teeth, which Cope believed to be weak and unable to manage any but the softest vegetation, turned out not to be so at all. While each individual tooth seemed feeble, when put in their correct places in the jaw they combined to form one of the most incredible plant-crushing and grinding arrangements known in the animal kingdom. In each jaw, hundreds of teeth can be seen. Each one appears as a small diamond-shaped shield, locked in against its neighbours by bony cement to form what is called a "battery" or "magazine" of teeth. They wore down from the top to provide a

119

LEFT: A "battery" of hundreds of diamond-shaped hadrosaur teeth.

The narrow "beak" of this duck-billed dinosaur was covered in life by a sharp horny beak for cropping tough plants. The long jaws were lined with hundreds of grinding teeth for pounding the toughest of food. The long, tubular crest on the head was thought by Professor Ostrom to contain the organ of smell.

THE HADROSAUR PARASAUROLOPHUS

ABOVE: The skeleton of this dinosaur is very well preserved and is typical of almost all hadrosaurs. The powerful hind legs were used for walking rather than swimming.

broad, rough pavement which acted like a millstone for grinding the toughest of plant food. Nor were the jaw muscles feeble. Ostrom was able to reconstruct the jaw muscles of these dinosaurs with some accuracy, and they clearly provided a very powerful bite.

It became clear that these animals were not necessarily bound to swamps for their existence; they could have fed upon the toughest of plants. In support of this, one of Sternberg's dinosaur mummies, which had been purchased by the Senckenberg Museum in Frankfurt, Germany, seemed to show evidence of a fossilized stomach (or at least something in the position where a stomach ought to be) and its contents. When examined these yielded conifer twigs, needles, seeds, and other fragments from land plants – not at all the soft vegetable diet previously imagined.

Not content to look at the feeding method in these animals, Ostrom also investigated the evidence for its sensory systems. The brain case revealed a fairly large brain, certainly large by reptile standards, indicating not only substantial intelligence, but also well developed senses. The eyes were large, suggesting good vision; the ears were well developed, indicating good hearing; and, most interestingly, there was evidence for a pretty good sense of smell – a faculty that is not well known among reptiles, apart from those which have a specialised forked tongue, which is used to pick up scent from the air. This led Ostrom into that mystery which had surrounded hadrosaurs from early this century: the function of the crest. Ostrom noticed that the crest was in the part of the air passage of the snout which is often associated with the olfactory membranes, those sensitive areas within the nose which detect smells and convey their messages directly back to the brain. He suggested therefore that the elaborate crests of hadrosaurs served as a housing for greatly enlarged olfactory membranes, which would have given these animals a most acute sense of smell.

A little later, he was to add to these observations on hadrosaurs from more extensive fieldwork, this time looking at footprints and the directions of trackways. He pointed out that a number of tracks showed significantly large numbers of individuals moving in the same direction at the same time. The implication that they moved in herds suggested two things: first that they were moving on land; and second that they were moving in social groups – with all the communication and interaction which that implies.

In the course of this work Ostrom was also able to question the belief that these animals were amphibious swamp dwellers. Anatomically they showed little evidence of persistent amphibious habits. True, the tail was deep, but it was stiffened by bony tendons which would have made it too stiff for efficient swimming. But a thick tail could be used as a counterbalance for walking and running on

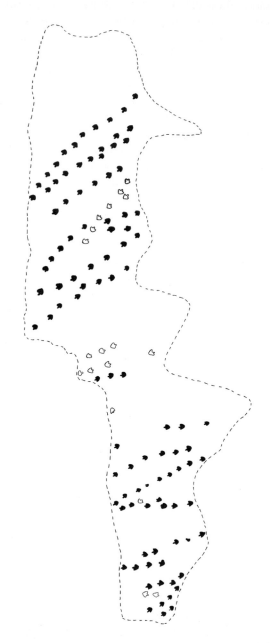

The consistently similar direction of the tracks of hadrosaurs (see the characteristically clover-leaf shape of the prints) is very strong evidence for herding and movement of these dinosaurs as socially integrated groups. This is one of the many pieces of evidence used by Professor John Ostrom.

land. The legs were long and powerful and the feet were rather narrow – unlike the broad, spreading foot that might be expected in an animal walking on soft mud all the time. The mitten of skin on the hand was open to question, but there was no real reason to suppose that these animals did not swim occasionally, in the same way that cattle or antelope do in order to cross rivers.

Although this is only a brief summary of a very lengthy piece of work, it does convey important messages about dinosaurs – messages which were, in a short number of years, to be reinforced by a most startling discovery with which John Ostrom was appropriately associated. Dinosaurs – or at least some types – had well developed senses and may have been social creatures which moved about in herds on land; they also possessed a sophisticated feeding system capable of dealing with the toughest of land plants.

The creature with terrible claws

During an expedition to southern Montana in 1964, Grant Meyer and John Ostrom discovered the first remains of a new Early Cretaceous carnivorous dinosaur. Theropod dinosaurs are rarely found, so any discovery is important; but this one seemed particularly interesting because it was obvious from early on that it was quite unlike other theropods. Over the next two field seasons the site was extensively excavated, and several hundred bones were recovered and taken back to Yale to be studied and described. Much painstaking work on this material has resulted in one of the best known and most interesting

The bony structure of the head of this Deinonychus *indicates that the skull was both light and strong. The hollow spaces allow for large muscles to attach to the jaws and produce a very strong bite, and the bony struts are arranged so that all the forces generated during the bite are efficiently transmitted from the bony edges of the jaws through to the firm area at the top of the skull.*

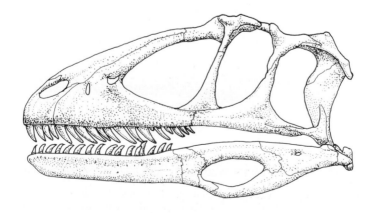

Armed with vicious teeth and claws, packs of Deinonychus *roamed North America during the Early Cretaceous pouncing on victims such as this large, but relatively harmless, ornithopod,* Iguanodon lakotaensis.

discoveries of any dinosaur. The animal was named *Deinonychus* ("terrible claw"), for reasons which will become obvious.

As I suggested in Chapter Two, theropods can usually be divided into two general types: those called carnosaurs which are big and have large heads, short, powerful necks, and short arms; and those called coelurosaurs which tend to be slender, have small heads, long slender necks, and long arms. *Deinonychus*, however, did not fit neatly into either of these categories. It was quite small – around 8 ft (2.4 m) long – yet it appeared to have a large head and strong neck combined with long grasping arms and fairly long legs. Added to this there were a number of other odd features which had never previously been recognized in theropod dinosaurs, and suggested that what had been discovered was something entirely new to the science.

The skull of *Deinonychus* is indeed large, compared with a typical coelurosaur, but it is also surprisingly light for its size, without sacrificing strength. There are a number of broad window-like openings in the side of the head, not only the obvious ones for its very large eyes but also others which are spanned by large jaw-closing muscles that would have given the beast a very strong bite. The jaws are long and lined with long, knife-like teeth with serrated edges. The teeth are also backward-pointing, which means that when the jaws closed on its prey, the more the prey struggled to pull itself free, the more it would drive the teeth into its own flesh – a hideously efficient design for the predator. Putting these observations together, the animal was clearly a sharp-eyed predator using its well armed, strongly muscled jaws to rapidly kill and eat its prey. It also combined, very efficiently, a large head and an ability to gulp large pieces of food, with the lightness essential for a fast-moving predator.

The neck was strong and flexible, giving the head a wide range of movement, and at the same time providing strength for the head to wrench large pieces of flesh from its unfortunate victim. By contrast the back was held quite stiff, so that it served as a rigid support for the weight of the chest and belly while the animal was walking or running. The chest also provided good anchorage for the strong muscles of the shoulders and arms.

The arms are long and seem to have been well muscled, with a long, three-fingered hand ending in viciously curved claws. These arms and hands were clearly ideally adapted for catching and holding on to struggling prey.

The legs were long and slender, as would be expected of a fast-running predator. That is not unusual; but the feet proved to be quite extraordinary. The theropod foot most often has three narrow, forward-pointing toes, and a much shorter backward-pointing toe, arranged like those of most birds. The foot of *Deinonychus* has quite a different arrangement. The first toe is very short and backward-pointed as expected, but the second toe bears a hugely enlarged, sickle-shaped claw (the "terrible claw" from which the animal gets its name), which is so large that the toe has to be hooked upward clear of the ground when the foot is placed on the ground. The third and fourth toes are long, slender, and forward-pointing, and roughly equal in length.

The tail, in outward appearance, would seem to be just a standard theropod tail – a long, slender counterbalance for the rest of the body while it is running on its hind legs. While this is generally true, there are some subtle differences to be found as well. The base of the tail has the expected row of rectangular vertebrae, but not far along the tail these become obscured by a sheath of long, thin bony rods running alongside the vertebrae. These rods are, in fact, hugely elongated extensions of what are normally small bony processes projecting from each vertebra. They would have made the end of the tail quite stiff, though its base would have been relatively mobile.

What does all this anatomical detail mean in relation to the animal's way of life? The image that emerged from Ostrom's research is one of a quite amazingly specialized predator. The head is clearly that of an alert predator; the eyes are very large, suggesting excellent vision, and the brain seems to have been large by reptilian standards, implying sharp senses and good coordination. But it is the body which really provides the most intriguing insights into the way of life of this dinosaur. The extraordinary foot with its enormously enlarged second claw was clearly a specialized weapon. A huge sickle could have been used for slashing by kicking movements of the leg. Although a hind foot might not seem the obvious place to keep a weapon of offence, in fact it does make sense if we look at the rest of the body. The long arms with their strong, grasping hands were obviously used to hold on to prey but, more than this, they may also have been used to position the prey ready for the death-dealing blows from the claws on the feet. *Deinonychus* would have held its prey at arms' length and kicked out at its soft, fleshy belly, brutally disembowelling it.

Catching the prey was also aided by some of the curious features of its anatomy. The legs were long and slender, as is typical of a fast runner. But the stiffened tail would also have played a role. When a bipedal animal runs, the tail can perform two vital functions. First and most obviously, it serves to balance the front end of the body. Second, it can have a much more dynamic function in increasing the animal's manoeuvrability. Flicking the tail to one side allows it to change direction, or "jink," very rapidly to catch the most evasive of prey. The combination of a flexible base and stiffening over most of its length would have allowed the tail to be swung quickly as a single unit, at the same time protecting it from damage by "whiplash." The stiffening would also have helped the tail to withstand the strong pull of the muscles that move the leg backwards at each stride.

As a relatively small predator by dinosaur standards this animal may well have selected smaller items of prey, such as small ornithopod dinosaurs, or perhaps any young dinosaurs that it same across. In addition – and this is in part inferred from the circumstances of its original discovery – it may have hunted in packs in the manner of a modern Cape hunting dog. Packs of hunting dogs are quite capable of bringing down prey considerably larger than themselves – such as full-grown antelope. It is quite rare to find large concentrations of theropod dinosaurs; they are nearly always represented by either isolated bones or a part skeleton of a single individual. The discovery of several hundred bones of *Deinonychus* in the Yale quarry in Montana suggests strongly that these dinosaurs may have roamed in packs, with their own social structure and hunting tactics.

The whole picture created by this animal is one of marvellous sophistication for its deadly way of life – in its own way no less marvellous than the lithe beauty and power of a lion. Does this really seem like a dull, slothful creature, lurching from one decaying carcass to the next, driven to move only by hunger – the model of gorgosaur envisaged by Lawrence Lambe earlier this century? The anatomy of *Deinonychus* requires a sophisticated sensory system and brain to coordinate its activities; it undoubtedly had good eyesight, and would have needed a well developed sense of balance and complex nervous and muscular coordination to have been able to catch fast-moving prey; and, once caught, to hold it down and kick it to death.

Put another way, the question could be: Do these characteristics fit with our understanding of the capabilities of modern reptiles? Are there any living reptiles they remotely resemble deinonychosaurs in their degree of anatomical sophistication? Again the answer seems to be No. Some of the very small lizards – such as geckos – are quite accomplished acrobats, but their abilities are greatly aided by their small size. Larger reptiles, including the big predators such as crocodiles and the larger monitor lizards, can be fast movers in very short bursts of activity. But these are low-slung, short-legged creatures, and in no way can their anatomy be compared adequately with that of the sleek and sophisticated deinonychosaur.

More than any other dinosaur *Deinonychus* focused attention on the paradoxical views which paleontologists seemed to have sought to undervalue the group as a whole. Not that there was ever an organized conspiracy to denigrate dinosaurs: it seems more the case that the great spectrum of known types simply allowed scientists with different views to find evidence to support their own pet theories about dinosaurs' ways of life.

DINOSAURS VERSUS MAMMALS

One fact that has been pondered over for many years has also been of great importance to the revival of the study of dinosaurs in recent years. That is that mammals are known to have lived at the time of the dinosaurs. The fact was known to Canon William Buckland (see page 69) in the early decades of the nineteenth century. By a remarkable coincidence, the tiny jaws of a mammal the size of a large shrew were found in the same rocks at the quarry in the village of Stonesfield, north of Oxford, in which the large bones of the first described dinosaur, *Megalosaurus*, were discovered.

In itself this discovery may not seem so remarkable, but the fact that mammals lived at the same time as dinosaurs does pose interesting questions. Today, mammals are extremely varied and successful creatures – particularly compared to modern reptiles. It therefore seems strange to imagine a world in which reptiles were dominant and mammals were apparently relegated to a minor role in all land ecosystems.

The reasons for the tremendous variety and dominance of mammals in many of the world's ecosystems are not too difficult to identify. Mammals are intelligent, resourceful, and highly adaptable creatures, as they need to be in an ever-changing environment. Mammals also have the ability to generate their own body heat internally by biochemical reactions – that is to say, they are "endotherms." They can control their body temperature very precisely, retaining heat by means of fur on the outside of the skin and fat just below it, or cooling themselves by panting or perspiring. In this way they can live independently of the prevailing environmental temperature, from the Equator to the Poles. They also have the advantage of giving birth to live young, and being able to nourish them on milk. The advantages of suckling are that the young can grow very quickly. During their growth they are protected and to some extent taught by their parents, so that they are best able to cope with an independent life.

There seems little doubt that the three factors of intelligence, endothermy, and parental care are the foundation upon which the success of modern mammals is built. All three certainly distinguish mammals from reptiles and can be seen to have positive advantages in a challenging environment. It would therefore seem natural to predict that when mammals first evolved on land they should have naturally replaced the less able or competent reptiles. However, a look at the fossil record rapidly dispels that notion.

The race starts

In the period immediately before the appearance of the first dinosaurs and mammals, during the Permian and Triassic Periods, the land had been dominated largely by the predecessors of true mammals. These animals are often known as "mammal-like reptiles." During this time they were not only very abundant (as we know from their plentiful fossil remains) but quite varied in type. Typical carnivorous types looked very like dogs, while the herbivores had the general appearance of modern pigs. Toward the end of the Triassic Period the appearance of these mammal-like reptiles came closer and closer to that of what we regard as true mammals, to the extent that the classificatory line drawn between the latest mammal-like reptile and the earliest mammal becomes almost arbitrary. The reason for this uncertainty is that the characteristics which define a true mammal are not ones which are likely to fossilise. Body temperature leaves no sign, nor does skin structure (particularly hair, or the soft non-scaly skin), nor is there evidence of live birth and suckling.

The earliest known fossils of creatures that are certainly mammals came from rocks in the Americas, Europe, South Africa, and China dating back to the time of transition between the Triassic and Jurassic Periods. These animals were small, shrew-like creatures – similar to the type which lived alongside *Megalosaurus* in the Late Jurassic. All these early mammals appear to have been small and nocturnal, probably living as scavengers and eating insects.

The earliest dinosaurs date back to a similar time. Late Triassic dinosaurs, the so-called "protodinosaurs" such as *Herrerasaurus* and *Staurikosaurus*, have been found in South America. However, these are very different in appearance from the early mammals, being relatively large predators, 3 to 8 ft (0.9-2.4 m) in size, with long, powerful legs.

So there were now two groups poised to take over. The mammal-like reptiles seem to have "prepared the ground" for the first true mammals, and it seems that their success should have been guaranteed. But things turned out otherwise.

Dinosaurs go ahead – for a while

Instead of the mammals, it was the dinosaurs which rapidly rose to dominate the land with a variety of large, powerful herbivores and carnivores. The mammals were relegated to a position of relative obscurity for the next 155 million years. The niche which they occupied for this enormous length of time, that of the nocturnal scavenger, suited them well. Small size gave them great agility and this, combined with their acute sensitivity and intelligence, allowed them to make the most of their lives. Small size also favors the development of endothermy, particularly when aided by the insulation of fur which protects the body from rapid changes in temperature.

It has been said that the mammals were "waiting in the wings" for their opportunity to expand and dominate terrestrial life, while the dinosaurs lorded it on land during the Mesozoic, That is undoubtedly true though, as far as the mammals living at the time were concerned, there was no choice but to wait. At the close of the Cretaceous Period, some 66 million years ago, all dinosaurs became extinct – for reasons that will be discussed later. With their disappearance the factors which had held the mammals in check in their nocturnal niches for so many million years were removed. The following era, the Cenozoic, saw the mammals rapidly rise to dominance.

Why did dinosaurs gain the lead?

The pattern in the fossil record showing the appearance of the earliest mammals at the same time as the dinosaurs and the subsequent success of dinosaurs, followed by the reversal of their fortunes 155 million years later, has been known to paleontologists for a considerable time. Natural theories attempting to explain this pattern and the puzzling failure of the mammals have been put forward, one of the first being Richard Owen's of 1842 (see page 74).

One of the most persuasive arguments used to explain the evident superiority of dinosaurs was developed by Dr Alan Charig of the Natural History Museum in London during the 1960s. Charig's starting point was the work of Cuvier and Owen. One of the key innovations in dinosaur design, when compared to that of standard reptiles, is to be found in the legs.

During the Triassic, Charig pointed out, archosaurs – ancestors of dinosaurs – showed a number of changes in the anatomy of their legs and hips, which can be explained as improvements to leg and muscle design. They became considerably less crocodile-like. Their heads are shorter and more compact, the front part of the body is shorter and more stocky – producing a shorter tail because it needs to counterbalance less weight at the hips – and the legs are significantly longer and have the possibility of being drawn in beneath the body, if not permanently at least while the animals are moving fast.

The critical changes can be traced to the hip socket and thigh bone (femur). The hip socket is cup-shaped and now has a bony rim along its upper edge, while the femur has become angled at the upper end so that its head points inwards. This arrangement allows the leg to be tucked in beneath the body as it is swung backward and forward during the animal's strides. While this is quite similar to the arrangement in a dinosaur, there are still important differences: the knee is still able to swivel, instead of acting as a simple hinge; the ankle is very complicated, again to allow a swivelling type of motion, rather than simple hinging; and finally the foot is still broad, rather than narrow as in dinosaurs. These differences in the lower leg joints allowed them to flex both when the legs were held straight down and when they were sprawled out sideways. In fact these Triassic archosaurs had a two-speed walking system: they could either sprawl their legs and crawl slowly along with the belly nearly dragging on the ground or, by tucking their legs in, they could adopt a "high walk" which would have been faster and more efficient.

Charig's proposal entered on the style of movement adopted by dinosaur ancestors. He suggested that these archosaurs with their two-speed legs became very successful toward the close of the Triassic Period. And indeed the fossil record does show that such creatures as *Ornithosuchus* and *Saurosuchus* were at this time some of the largest and most aggressive carnivores. The efficiency of their legs enabled these animals to capture prey more easily. The result was a rapid decline in the abundance of other groups, such as the mammal-like reptiles which, he agreed, had not perfected their running to the same degree as their competitors. Ultimately, so the argument goes, the mammal-like reptiles became extinct under this predator pressure, leaving the archosaurs and early dinosaurs to continue.

This argument can clearly be seen as an example of the famous dictum, "The race to the swift, the battle to the strong." The dinosaurs' ancestors were both swift and strong, and they prevailed. Dinosaurs, then, are seen as a simple extension of an evolutionary trend started by their predecessors. They perfected their leg mechanics to the point where they no longer needed the slow sprawl. Having achieved this level of perfection, the pillar-like legs were used by dinosaurs to maximum advantage in the evolution of a vast range of types, from very swift-moving and lightly built forms with long-striding legs, especially the theropods, to gigantic "earth shakers" with stout, pillar-like legs, notably the sauropods.

Warm-blooded dinosaurs?

During the late 1960s one of Professor Ostrom's students at Yale, Robert Bakker, was also looking at the question of the dinosaurs' success in the Triassic. Bakker took several strands from discussions and theories relating to dinosaurs, added ideas of his own, and put forward a new general theory for the success of dinosaurs. This was that dinosaurs must have been "warm-blooded" – by which he meant endothermic. This implied that they were much more similar to living mammals and birds in their physiology than to modern reptiles.

Bakker's argument is deceptively simple and goes back to the observations we have just been discussing concerning the fossil history of mammals and dinosaurs. We know that dinosaurs and early mammals appeared at roughly the same time. Our modern knowledge of the attributes of mammals leads us to expect that, if dinosaurs

131

were conventional – that is, ectothermic – reptiles, mammals should have risen to dominate terrestrial ecosystems 155 million years earlier than they actually did. For dinosaurs to have dominated mammals for such a long period it must mean that they, as a group, had biological characteristics which were at least as sophisticated as those of the mammals. Since the brains, and consequently the intelligence, of early mammals and dinosaurs do not seem to have wildly different, and the extent to which either of them nurtured their young is debatable, the key feature must have been endothermic physiology.

Starting with the assumption that dinosaurs were endothermic, Bakker set himself the task of discovering whether other evidence from dinosaur anatomy and from other clues in the fossil record supported this assumption. He began developing arguments and arranging data to support his view.

One argument was a simple one of anatomy. It is well known that dinosaurs had upright legs and walked in the same way as mammals and birds. No other vertebrates are capable of walking in this way at the present time. Since both mammals and birds are endothermic, surely this was conclusive evidence that upright legs and endothermy go together. Endothermy, by maintaining temperature at an ideal level, allows an animal to deliver energy to all its systems as efficiently as possible. In simple terms, it has a more powerful "engine." Since the design of a dinosaur's leg is one that would allow it to run fast, it seems unlikely that Nature would devise such a system if the "engine" driving the legs could not deliver the power necessary to allow the dinosaur to move fast.

Bakker's famous illustration of an alert, dynamic Deinonychus *sums up much of Ostrom's work: it seems that some of the smaller, livelier dinosaurs were closer to mammals and birds than modern reptiles in their levels of activity - but were they "warm-blooded"? The evidence is still finely balanced.*

Bakker '69

Still considering posture, Bakker also looked carefully at the anatomy of Marsh's sauropods, and the long-held view that they lived in swamps. Rather than finding features associated with swamp dwellers, all he found were ones consistent with animals living most of their lives on land. For example the legs were pillar-like, with hollow bones for lightness combined with strength. Also the feet were very narrow in relation to the size of the animal, rather like those of an elephant, and clearly adapted for walking on firm ground rather than in sticky mud. Similarly the chest was not rounded and barrel-shaped as in the amphibious hippopotamus, but narrow and deep, as in the elephant, giving support to the chest and belly under the effect of gravity on land rather than in water. Finally, the fact that the bony openings for the nostrils were perched on top of the head was not a compelling reason for believing that these animals lived in water, for a surprising range of creatures – including elephants and tapirs have high-positioned nostrils, and no one could argue that this was because they lived in water. This suggested a much more active life for sauropods as high arboreal browsers; one that had previously been proposed toward the end of the last century by Cope (see page 96). Not only was a giraffe-style mode of life reinstated, but Bakker went further by suggesting that enormous sauropods such as *Diplodocus* may have been able to rear up on their hind legs to reach higher foliage.

Another of his arguments concerned the ratios of the number of predators to that of prey animals. He suggested a completely novel idea for assessing the physiology of fossil creatures: that of taking cen-

suses of fossil collections. His reasoning was as follows. Endothermy has one very noticeable cost when ectotherms and endotherms are compared: endotherms, which use energy to keep warm, therefore have much larger appetites! Taking very approximate figures, if a crocodile and a lion of equal body weight are compared, the lion will eat about ten times as much as a crocodile in any given period of time. The question is: Can this difference in appetite be measured in the fossil record? Bakker claimed that it could.

He analyzed the numbers of recorded fossils of predators and prey animals from the Palaeozoic (Permian Age collections containing primitive reptiles and amphibians), Mesozoic (dinosaur collections), and Censozoic (mammal collections). The results, he claimed, showed that there were roughly equal numbers of predators and prey in the Palaeozoic collections (that is, the predator–prey ratio was approximately one to one) whereas the collections of both dinosaur and mammal fossils showed only one predator to more than ten prey.

His conclusion was that the ancient reptiles and amphibians showed roughly equal numbers of predators and prey because the predators were ectothermic and fed infrequently. By contrast mammals, which are known as endotherms, need far more prey animals because the predators feed ten times as frequently as ectotherms. Since the results from the census of the dinosaur collections roughly matched that of the mammal collections, he claimed this as evidence that dinosaurs were endothermic as well.

Bakker also examined the structure of fossil bones. If very thin sections of bone are cut and polished until they are transparent, the fine structure of the bone can be seen in detail under a microscope. Bone structure can frequently reveal the conditions under which the animal lived. Slow growth, for example, may result in very dense bone with few if any channels running through it carrying blood vessels. Fast-growing bone, which is associated with highly active animals, is likely to have many channels for blood vessels.

Crude comparisons between polished sections of lizard and cow bones show a huge difference in the quantity of blood vessels running through them. Dinosaur bones when sectioned seem to show an arrangement much more similar to that of a typical mammal. Again the conclusion based on this observation is that the dinosaur and the mammal are much more similar in their physiology than a dinosaur is to a modern reptile – another confirmation of dinosaurs being endothermic.

Feeding mechanisms provided another argument. Animals which need to feed regularly because they are endotherms would be expected to have complex jaws and teeth for processing large quantities of food. Again dinosaurs seem to conform to this model. In the case of hadrosaurs it is clearly the case that their beaks were very effective croppers, and their huge dental batteries with hundreds of teeth in each jaw were able to pound up the toughest vegetation.

The sauropods seem at first sight to be an exception because they are the biggest dinosaurs, with perhaps the biggest appetites, and yet they have the most feeble of jaws and teeth. However, there is a ready explanation for this, which is that the teeth simply served as a cropping device, and were not for chewing – in fact they were the equivalent of the hadrosaur's beak. The comb-like arrangement of teeth could literally rake in leaves and twigs which were swallowed and sent to the huge, muscular stomach, where there was a stone-lined gizzard like that of a bird, which pounded up the vegetation into a digestible soup. Here again, dinosaurs seem to fit his expectations very well.

Other scientists' theories could be fitted with Bakker's – for example the evidence from John Ostrom's work on hadrosaurs which suggested that they were social creatures living in herds. This implied that at least these types of dinosaurs engaged in a variety of social behavior necessary to keep a herd together. Further work on sauropod footprints was interpreted as indicating herding in these creatures as well. In fact Bakker went so far as to suggest that sauropods may well have moved around in highly structured herds. Small, younger individuals and females moved along in the center of the herd, while the larger males patrolled the fringes of the herd as protection against marauding allosaurs.

In this instance the inference about the physiology of dinosaurs is a little more subtle. First it emphasizes that social interactions were complex in dinosaur communities, and then asks whether such behavior is normally associated with modern ectothermic reptiles or with endothermic mammals and birds.

Over the years there has been much dispute about Bakker's analyses and conclusions. But this debate has been valuable. Research by Professor James Hopson at Chicago has shown that, by the standard of reptiles, most dinosaurs had a large brain cavity.

Some other extremely timely and appropriate work had come from Professor Ostrom, with his demonstration that there is now an almost indisputable link between dinosaurs and birds (see page 191). Allying dinosaurs with a group that is demonstrably endothermic adds further weight to Bakker's argument for warm-blooded dinosaurs.

Whatever the rights and wrongs of Bakker's arguments, and his particular approach to paleontology as a science, his contributions, when allied to the more sober and reflective work of others – of which the most notable is Professor Ostrom – have had a galvanizing effect on dinosaur paleontology in recent years.

Trackway evidence suggests that large sauropods moved about in tightly knit and highly structured herds, with adults on the outside to protect their smaller and vulnerable young from marauding predators. These allosaurs risk being crushed by 20-30 tons of sauropod foot.

ABOVE: *Middle Triassic faunas (from the top):* Ticinosuchus, *a typical archosaur;* Kannemeyeria, *a dicynodont; the carnivorous cynodont* Cynognathus; *the rhynchosaur* Scaphonyx *and* Massetognathus, *a smaller relative of* Cynognathus.

RIGHT: *Large, rapacious claws like this one on the hand of* Allosaurus *proved to be important in the new interpretations that were to emerge during the late 1950s and 1970s as dinosaurs came to be seen as very active, dynamic creatures.*

CHAPTER FIVE

Late News from the Mesozoic

A great deal of research into dinosaurs is currently going on – far too much to allow more than an outline to be given here. But I hope that I will be able to give at least an impression of the variety of work being done.

The origin of dinosaurs was a crucial event in the unfolding of life in the Mesozoic Era, marking a rapid change in the fortunes of a number of different groups of animals. Various researchers have been studying Late Triassic rocks in different parts of the world, the configuration of the continents at the time, the environment, and the plants and animals of the time, in an attempt to piece together a clearer picture of the circumstances which led to the rise of the dinosaurs and the decline of so many other groups. They hope to find some common factor here which may help to explain why dinosaurs survived while others fail.

Before the dinosaurs

Work on Middle and Late Triassic faunas around the world has shown a great variety of animal types coexisting. During the Middle Triassic there appear to have been a wide range of mammal-like reptiles, most of which are of only moderate size, 1 to 4 ft (0.3-1.2 m) long. The mainly herbivorous dicynodonts were characterized by a turtle-like horny beak at the front of the snout, and two large, tusk-like teeth projecting down from the sides of the upper jaw (their name means "two dog teeth"). Despite their lack of other teeth these animals were very

139

By the Late Triassic archosaurs begin to dominate, such as the large, agile carnivores Ornithosuchus *and* Saurosuchus *(top), and the medium-sized, heavily armored, snub-nosed plant-eating archosaur, the aetosaur* Stagonolepis. *An early flying archosaur, the pterosaur* Eudimorphodon *is seen being pursued by the diminutive, very dinosaur-like* Lagosuchus. *At the bottom right can be seen the small remnants of the mammal-like reptile lineage,* Oligokyphus *(left), and* Trithelodon *(right).*

effective herbivores, cutting and grinding plant food with their sharp beaks, mobile jaws and very powerful jaw muscles. These rather stout, barrel-bodied animals had quite short tails, but moderately long legs which they were able to tuck beneath their bodies when they wished to move more quickly. In fact they seem to have developed much the same two-speed system for moving their legs as was described for the archosaurs in the previous chapter (see page 130).

Living alongside these types were other somewhat more slender-bodied herbivorous mammal-like reptiles. Unlike the dicynodonts, these others had elaborate cheek teeth which allowed them to chew and crush up plants very effectively. As Triassic times passed the dicynodonts seem to have declined in number and variety in favor of the toothed herbivores.

The carnivorous counterparts of these plant eaters were known as cynodonts ("dog teeth"). Their teeth and jaws were remarkably similar in shape to those of modern dogs – hence their name and implied predatory habits. They had slim, dog-like bodies and relatively long legs again capable of being held in a sprawling position or beneath the body for fast running. However they do not seem to have been capable of using the bounding run or gallop of a modern dog, because the spine was capable only of flexing from side to side, not up and down. They would have waddled as they ran.

Living alongside these groups were the archosaurs, which must have included among their ranks genuine ancestors of the dinosaurs. During the Middle Triassic many of these creatures were fairly large – up to about 14 ft (4.2 m) long. Most of them had the appearance of long-legged, snub-nosed crocodiles; *Ticinosuchus* is an example. But, unlike crocodiles, these appear to have been completely terrestrial creatures – hence the long-striding legs. They had deep skulls whose jaws were lined with narrow, knife-shaped teeth. These animals held their bodies quite stiffly, as we know from the limited range of movement of the rows of bony scales down the back, so that when they ran the body tended to waddle very little as the legs swung to and fro beneath. The two-speed legs of these creatures were similar to those of the mammal-like reptiles that they undoubtedly preyed upon, but they could doubtless cover the ground more quickly with their longer legs.

Another group which made its first appearance in the Middle Triassic was the rhynchosaurs ("snouted lizards"). Outwardly these creatures resembled dicynodonts, with similar barrel bodies, and they too were herbivores. Their heads were their most distinctive feature: generally very broad at the back, with high-placed eyes and a downward-pointing snout which tapered to a curiously hooked beak made of bone. Their jaws were lined with fields of pimple-like teeth which formed grinding plates for the crushing of plant food.

141

In Late Triassic times a whole host of new groups appeared and various others disappeared. Of the groups we have already met, the dicynodonts had virtually disappeared, having been replaced by smaller herbivorous mammal-like reptiles, and very probably by the rhynchosaurs which became abundant for a short while in the latest Triassic. The carnivorous cynodonts remained fairly numerous, but they too evolved towards smaller, more agile types. Right at the close of this period the first true mammals must have had their origins, appearing as small insectivores (insect eaters) from ancestors which seem to be traceable back through the smaller Late Triassic cynodonts.

The archosaurs, by contrast, diversified quite considerably. The earliest true crocodiles appear during this time, as well as another group of very crocodile-like archosaurs known as phytosaurs ("plant lizards"), distinguished from crocodiles by having their nostrils placed on a mound just in front of their eyes, rather than on the tip of the snout as in true crocodiles. Another group of archosaurs, whose history is mysterious, appears to have developed into small, probably tree-living insectivores which eventually achieved the power of flight: these were the early pterosaurs ("winged lizards"). It is possible that their origins may well date back somewhat earlier, but because they lived in trees and were very small, their remains are very rarely fossilized.

On the land the archosaurs also produced a variety of carnivorous forms. Some were similar to the animals seen in the Middle Triassic but larger and more powerful, an example being *Saurosuchus*. Others had enlarged back legs and tended to run on these rather than on all fours. These included quite large, heavy-bodied creatures, such as *Ornithosuchus*; but others, such as *Lagosuchus*, were of a much more lightly built variety and in many ways seem to be anticipating the design of the earliest dinosaurs. Their legs are very long, and were clearly held beneath the body for much of the time; they were agile, fast-moving bipedal predators.

In addition to the carnivorous varieties of archosaur, there were also some new herbivorous types. These include the aetosaurs ("armored lizards") such as *Stagonolepis*, which have blunt snouts and chipping teeth, and heavy armor.

True dinosaurs make their first appearance during the Late Triassic. Some of the earliest ones known have been found in Argentina. The best known examples are *Staurikosaurus*, a rather slender carnivore, known from fragments only; and *Herrerasaurus*, a larger and more heavily built dinosaur which is known from several partial skeletons. In recent years a number of expeditions arranged by Dr José Bonaparte of the Museum of La Plata in Buenos Aires, with collaboration from Dr Paul Sereno of the University of Chicago, have explored several sites for herrerasaur material and discovered several new and important skeletons, including that of another very early new carnivorous dinosaur which has been named *Eoraptor* ("dawn predator").

A little later on in the Triassic, other dinosaur types are found in various parts of the world. These show that, once the dinosaurs had appeared, they rapidly began to diversify. *Coelophysis* and *Syntarsus* are two slender, fast-moving theropods found very close to the end of the Triassic and into the Early Jurassic. But another equally important group was that of the early sauropodomorphs, which now appear right across the globe. These are represented both by slender, plant-eating creatures such as *Massospondylus* and *Anchisaurus*, known from North American and Southern Africa, and by larger and heavier forms such as *Plateosaurus* from Germany, *Lufengosaurus* from China, and *Riojasaurus* from South America. Curiously enough, these are the first high arboreal browsers that ever existed on Earth. They seem to have been very abundant, judging by the number of remains that have been found to date.

Triassic climate and plants

During the Late Triassic the continents were largely joined together to form the supercontinent Pangaea. This can be confirmed in various ways, for example paleomagnetism: the natural magnetism of rocks shows how they were aligned in relation to the Earth's magnetic field when they were formed, and thus how the whole continent of which they are a part was aligned at that time. It is also shown by the fact that the animals living at this time seem to be widely distributed across the world. For example, the early dinosaur *Coelophysis* is found in North America; a very close relative, *Syntarsus*, has been found in southern Africa; and footprints which may belong to a *Coelophysis*-like dinosaur have been found in Britain too, in South Wales.

The environment of the Late Triassic was undoubtedly governed by the supercontinental landmass, which would have produced reasonably humid conditions near the coast but a very arid interior. As we have already seen (page 40), this was a time of general global warmth compared with today, and also of considerable seasonal aridity and widespread desert or semi-arid conditions. In the lowland, better watered areas and around the edge of the continent, plant life was dominated by ferns covering the ground, tree-like seed ferns such as *Dicroidium*, and conifers and cycads adapted to a fairly dry climate.

This global picture is, however, still rather crude. But from recent studies by paleobotanists it is clear that there are differences in the types of plants predominating in the northern and southern areas of Pangaea, which suggests that there may have been differences in the general environments.

More ideas on dinosaur origins

We have already discussed the two theories of Charig and Bakker concerning the origin of dinosaurs in the Late Triassic and their subsequent success. The first idea was that the mechanics of their legs gave them a crucial advantage over their contemporaries. The second was that they had an endothermic physiology – that is, they were "warm-blooded" – and thus on an equal footing with the mammals which had appeared at about the same time. These are not the only theories.

Coping with drought

A particularly interesting theory was put forward by Dr Pamela Robinson of University College, London. The Late Triassic is thought to have been a time of predominantly hot, arid conditions worldwide. What kinds of animal are best adapted to survive in such conditions? If we are comparing reptiles and mammals – which is the object of the exercise – then the answer would appear to be reptiles.

Reptiles have tough, scaly skins which protect them from drying out. They excrete very little water: rather than losing a lot of liquid as urine, they produce a whitish, paste-like substance similar to bird droppings. Because they are ectothermic ("cold-blooded") – which is certainly true of modern reptiles – they do not expend energy keeping themselves warm, and therefore eat little and can thrive where food is scarce and of poor quality.

By contrast mammals have soft, flexible skins through which water can pass easily. They lose a great deal of water in their urine (though some modern desert mammals have become very economical in this respect). They are endothermic, which has two major drawbacks in hot desert conditions: to keep themselves cool they must sweat or pant, both of which cause them to lose much water; and because they are endothermic they need plentiful quantities of food. All these attributes count against mammals in hot deserts, where water and food are scarce and the sun can cause rapid overheating and heavy water loss.

Survivors of a catastrophe

An altogether different proposal has been put forward by another British scientist, Dr Michael Benton of Bristol University. He developed this from the ideas of paleobiologists in America (notably Professors Stephen Jay Gould, Steven Stanley, Jack Sepkoski, and David Raup) who have been working on the evolution of clams and other marine groups. These ideas revolve around the understanding that many of the large and sudden changes in the history of life on Earth can be explained as resulting from some kind of external disturbance to the environment. Detailed analysis of the time of appearance and disappearance of animals as the Triassic gave way to the Jurassic has suggested to Benton that the success of the dinosaurs was linked not to some specific adaptive feature such as limb mechanics, or endothermy or ectothermy, but to a catastrophic event which caused the mass extinction of a wide range of land-living animals and the fortunate survival of dinosaurs and a few other groups.

Who is right?

How do we choose between these competing theories? The simple answer is that we cannot. The theories based on the supposed physical nature of dinosaurs cannot be properly tested, because we can never know exactly how dinosaurs, their predecessors, or their contemporaries functioned. The catastrophe theory may be confirmed or weakened by new discoveries.

That may sound pessimistic, but in fact such doubt is beneficial. Competing theories spur on discovery and creative thinking. At the moment various teams are looking at precisely these problems, trying to discover more about the distribution of fossils across the Triassic–Jurassic boundary in various parts of the world.

Some of that work is being done in Nova Scotia by Dr Paul Olsen and Dr Neil Shubin of the Lamont-Doherty Observatory. The results are most intriguing. He and his colleagues have been able to chart the diversity of animals and plants whose remains were washed into a permanent lake. The lake deposits show a mixed fauna of early dinosaurs and mammal-like reptiles along with a variety of other types evolving together up to the end of the Triassic. There is then a fairly abrupt change where Triassic forms disappear and give way to a much less varied selection of small early dinosaurs, small mammal-like reptiles and small crocodiles.

So, at least in the area that is now Nova Scotia, dinosaurs seem to have arisen and shared their world with a wide range of animal types as part of a mixed community, in the absence of true mammals. Some major event seems to have taken place which swept away most of the early dinosaurs as well as many other groups, leaving only small animals as survivors. If that is not simply a local phenomenon relating to the lake and its immediate environment, but can genuinely be translated to a worldwide event, it may be extremely important. It requires an explanation which connects the evolution of mixed communities which include early dinosaurs with a later mass extinction from which only small animals survived.

Clearly matters are far from settled, and a great deal of research and interpretation remains to be done before we will be any closer to a consensus on the origin of dinosaurs.

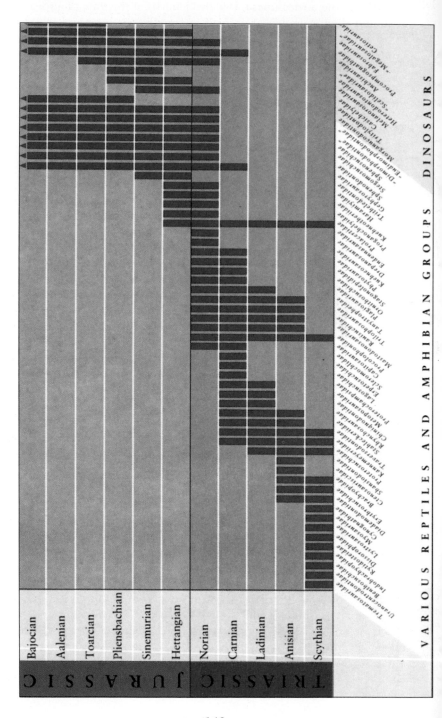

VARIOUS REPTILES AND AMPHIBIAN GROUPS DINOSAURS

DINOSAUR LEGS

Sacral ribs fix ilium
to vertebrae

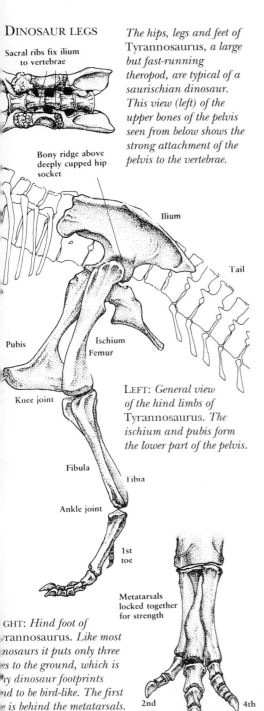

Bony ridge above
deeply cupped hip
socket

The hips, legs and feet of
Tyrannosaurus, *a large
but fast-running
theropod, are typical of a
saurischian dinosaur.
This view (left) of the
upper bones of the pelvis
seen from below shows the
strong attachment of the
pelvis to the vertebrae.*

Ilium

Tail

Pubis

Ischium
Femur

Knee joint

LEFT: *General view
of the hind limbs of*
Tyrannosaurus. *The
ischium and pubis form
the lower part of the pelvis.*

Fibula

Tibia

Ankle joint

1st
toe

Metatarsals
locked together
for strength

GHT: *Hind foot of
rannosaurus. Like most
nosaurs it puts only three
es to the ground, which is
y dinosaur footprints
nd to be bird-like. The first
e is behind the metatarsals.
here is no fifth toe.*

2nd
toe

4th
toe

3rd toe

LOOKING AT LEGS

As mentioned earlier, one of
the chief characteristics of
dinosaurs is that they walk with
their legs tucked neatly
beneath their bodies. The
pillar-like arrangement of the
leg bones, which made them
resemble the legs of elephants,
was a feature originally noted
by Owen. It has interesting consequences for the shape of all
the bones and joints of the leg.

The hip bones are very
rigidly attached to the backbone, by means of short, strong
ribs of which there are usually
four or five on each side. The
hip socket is deeply cupped to
form a firm attachment for the
femur, or upper leg bone.

The femur is shaped so that
it will swing underneath the
body. Its upper end, which fits
snugly in to the hip socket, is
bent sharply inward and has a
smooth ball-shaped top so that
the bone can pivot freely forward and back. The main
length of the bone is more or
less straight, though complicated here and there by bumps
and ridges where the leg muscles are attached. At the lower
end of this bone there is a
smooth bearing surface for the
knee joint, which is shaped in
such a way that the knee will
only flex back and forth – just
like the human knee joint.

The lower leg consists of two
bones lying side by side. Both
are almost straight. The upper

147

ends form a simple knee hinge with the femur, and the lower ends are capped by two ankle bones. The latter are very firmly attached to the lower leg, and form the upper half of the main ankle joint, which again is simple and allows the foot to swing to and fro. The human ankle joint is quite loose compared to that of a dinosaur, so a better example is the ankle of a horse, which is very limited and is unable to make any sideways movement.

A dinosaur's foot is quite different from a human one. The long bones (metatarsals) are not nearly parallel with the ground, but are bunched together and run upwards to the ankle joint, so that dinosaurs walk on their toes at all times. The toes tend to be quite long and slender. This gives the foot a firm grip of the ground and helps the dinosaur to balance. The long toes also increase the creature's stride length – and therefore the speed at which it walks or runs – and gives a characteristically narrow foot shape. Nearly all dinosaurs have only three walking toes on the foot.

There are of course exceptions to every rule, and some of the very heavy dinosaurs, for example the giant sauropods *Diplodocus* and *Brachiosaurus*, have shorter and broader toes, rather like those of an elephant. But these still show their origins from earlier, narrower, and more typical dinosaur feet.

DINOSAUR POSTURE

Dinosaur postures fall into two quite distinct categories, as we have seen from the dispute between Owen with his "elephantine reptiles," and Leidy and Huxley with their kangaroo – or bird-like models for dinosaurs.

On two legs

The earliest dinosaurs known are bipedal, and this may well be the original posture of the entire group. In this pose, the hind legs are long and relatively slender, giving a long-striding gait which can permit fast running. Since these animals are not heavily built, the legs can be light and delicate. More often than not the lower leg bones, notably those of the shin (the tibia and fibula), and the metatarsals which form the upper part of the foot to which the toes are attached are noticeably elongated, which improves the stride length at the expense of some overall strength. The feet are also narrow, and the toes slender, consisting of just three forward-pointing toes (numbers 2, 3, and 4) and a very much reduced, or completely absent first toe. The foot is also held in what is called a digitigrade (literally "toe-walking") position, which means that the long bones of the foot (metatarsals) are held permanently clear of the ground and the ani-

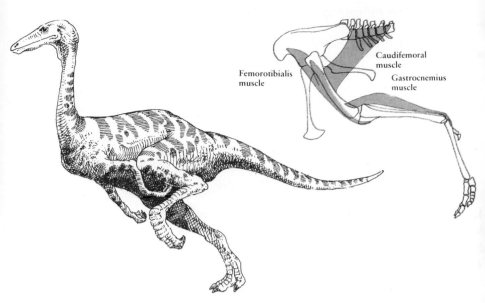

Femorotibialis
muscle

Caudifemoral
muscle

Gastrocnemius
muscle

This restoration of an ornithomimid theropod shows the delicacy and poise which contributed to the effectiveness of these creatures. The source of power for running came from the main leg and hip muscles shown above right. This power was generated by the caudifemoral muscle, while the femorotibialis (thigh muscles) and gastrocnemius (calf muscles) act to straighten the leg giving a powerful push.

mal walks only on its toes. Such an arrangement increases the effective length of the leg, and also saves energy because it means that the body does not have to be raised and lowered every time the foot is lifted.

The muscles which operate the leg are large and powerful, and are clustered toward the top of the leg – as is usual in fast-running animals, which need to be light-footed. The most powerful muscles are those which run from the back of the upper leg onto the side of the tail (caudifemoral muscles), and pull the leg back, driving the body forward with each stride; the large thigh muscles (femorotibialis) which straighten the knee; and the calf muscles (gastrocnemius) which extend the foot.

The balance of the body at the hips is made possible by the large tail which sticks out from the hips as a counterweight. In front of the hips, the chest tends to be fairly compact, and the neck is bent quite sharply up. Both mean that the body does not project very far forwards, and thus that the tail does not have to be enormously heavy to balance them. Flexing the neck also raises the head, giving it a high vantage point, so it is not surprising that the eyes are large and well-developed. Indeed judging from the relatively forward position of their eye sockets, some theropods may have even had stereoscopic

THEROPOD POSTURE
The powerful legs supported the weight of large theropods such as this tyrannosaur and the backbone acted like a flexible yet strong girder to support the body. The gastralia lining the belly can clearly be seen in this diagram.

Not all theropods are large but they are all bipedal, running on their hind legs and using their long tails as a counterbalance.

This gives them great agility when hunting, leaving their hands free for catching prey.

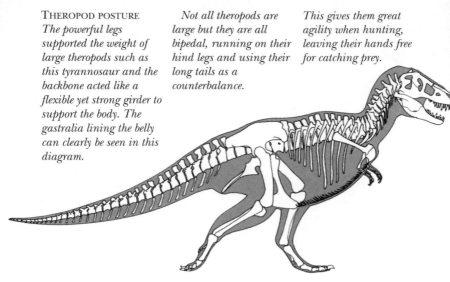

vision – that is, they could see an object with both eyes at once, allowing them to judge its distance accurately. The chest and belly are slung beneath the backbone; and the front limbs, freed from walking duties, develop into grasping hands.

This general type of body plan is seen in a range of carnivores, including all the theropods; as well as in herbivores, for it is also the standard ornithopod design. There are, however, subtle differences between the theropod and the ornithopod.

Bipedal theropods The backbone of theropods has to act like a rigid beam supporting the entire weight of the body over the hips, and it is variously modified to strengthen it. The tail is sometimes stiffened by bony rods which partly ensheath the tail bones; this is necessary to withstand the forces exerted upon it by the large retractor muscles pulling the leg back, as well as to prevent whiplash problems when it is being used as a dynamic stabilizer (see page 255).

The hips are joined to the backbone by very short, strong ribs, known as sacral ribs. The vertebrae forward of the hips have large, square spines which project upwards between the muscles of the back. These spines strengthen the backbone to prevent it from sagging under the weight of the belly and chest, by means of sheets of strong ligaments stretched between the front and rear edges of adjacent spines. In the larger theropods such as *Tyrannosaurus* the vertebrae are often perforated by large openings and spaces in their sides. These openings, called pleurocoels, help to save weight in these large animals, and may also be associated with an air sac system such as that found in modern birds, consisting of a series of air-filled passages connected to the lungs.

150

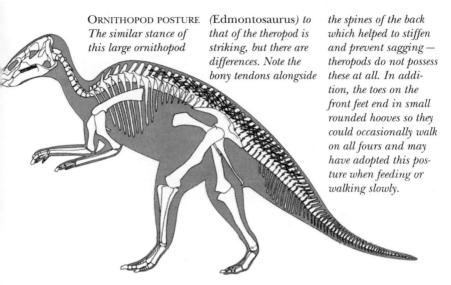

ORNITHOPOD POSTURE *The similar stance of this large ornithopod* (Edmontosaurus) *to that of the theropod is striking, but there are differences. Note the bony tendons alongside* *the spines of the back which helped to stiffen and prevent sagging — theropods do not possess these at all. In addition, the toes on the front feet end in small rounded hooves so they could occasionally walk on all fours and may have adopted this posture when feeding or walking slowly.*

The rigid backbone is balanced by an arrangement of muscles and bones which line the belly of theropods. The bones, known as gastralia or "belly ribs," run from the shoulders right down to low hip level. They served to protect and strengthen the belly region, rather like a permanently fixed bodice.

The front limbs of theropods are variously modified. In most the hands are three-fingered and provided with sharp claws for seizing prey. Some, however, such as the ornithomimids, tend to develop narrow fingers with less sharply curved claws, which were clearly used for careful manipulation of objects. Tyrannosaurs, despite their huge size, seem to have surprisingly short arms with only two claws, and a similar arrangement is found independently in the South American theropod *Carnotaurus* which has forearms shortened almost to nothing. The newly discovered theropod from Britain, *Baryonyx*, has surprisingly large fore limbs, the hand of which may well have been equipped with a large, gaff-like claw (see page 158).

Bipedal ornithopods The horizontal pose of the backbone of ornithopods presents exactly the same requirements for strengthening as that of theropods. The ornithopod solution is, however, totally different. Ornithopods tend to have rather longer spines on their vertebrae both in the tail and back areas. Normally there would be large muscles on either side of the spines serving to strengthen and stiffen the back, as appears to have been the case in theropods. However, ornithopods seem to have drastically modified these back muscles and converted them into long, bony "ossified tendons" (see page 94). These are arranged into a trelliswork of long rods which supported the back without the need for heavy (and energy-sapping) back muscles.

151

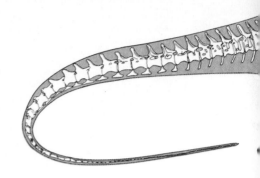

DIPLODOCUS POSTURE
The huge digestive tract of the larger sauropods pushes the body weight forward, requiring them to spread their weight on four pillar-like legs. The vertebrae have hollow areas (pleurocoels) to help save weight without affecting the strength of the skeleton. The neck could be craned up to reach tree top foliage.

The structure of the suspension bridge is surprisingly similar to the large sauropod dinosaur.

The belly in plant eaters tends to be much larger than in meat eaters. In ornithischians the pubis bone – the bone at the lower front of the pelvis – is rotated backwards to lie against the ischium – the bone at the back. By doing this a great deal more space is left for the belly, which is free to extend back between the legs. This is doubly advantageous for ornithopods, because it not only allows them a large belly but also permits them to remain agile bipeds, because the weight of the stomach is shifted back between the legs so that the animal can retain its balance. Most often a large belly obliges an animal to walk on all fours. The belly does not appear to have been lined with gastralia in ornithopods, and this may be associated with the need for a larger and more flexible belly in these animals. Unlike the large theropods, there is no evidence of pleurocoels in the sides of the vertebrae in even the largest ornithopods – some of which, such as *Edmontosaurus*, *Shantungosaurus*, and *Saurolophus* reached lengths of 44 ft (13.4 m).

The fore limbs and hands of ornithopods are not as varied as those of theropods. In the smaller varieties such as *Hypsilophodon*, *Heterodontosaurus*, *Dryosaurus* and *Orodromeus* the hand is broad and five-fingered, with short, blunt-clawed fingers used for grasping. In larger types like *Iguanodon*, *Ouranosaurus*, and the hadrosaurs the hand is longer and the middle fingers develop broad hooves instead of claws, which were clearly used for walking with.

On four legs

Quadrupedal dinosaurs appear a little later in the fossil record than the bipedal dinosaurs, but are equally interesting and considerably more varied. Almost all quadrupedal dinosaurs are herbivores. This is accounted for by the fact that the stomachs of plant eaters need to be very large and, unless the body is specially modified as in ornithopods, this inevitably pitches the body forwards onto its fore limbs.

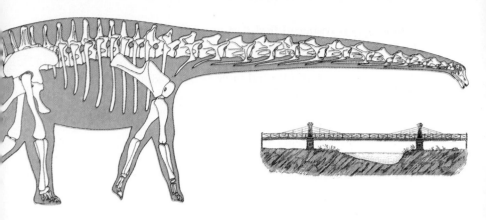

Apart from a few of the smaller types, nearly all quadrupedal dinosaurs have very stout pillar-like legs, which are designed to carry great weight, rather than to move the animal swiftly. As a result of this, the proportions of the bones of the leg tend to be the exact reverse of those of the biped. The thigh bone (femur) is the longest and is very straight, as bends in the shaft would tend to weaken it; the bones of the shin (tibia and fibula) are shorter than the femur; and the bones of the foot are very short, ending in a broad set of stubby toes. The arrangement of the bones of the foot is remarkably similar to that of a living elephant and, again in common with an elephant, the leg was probably kept more or less straight during a normal stride. The foot of the quadruped is particularly interesting because of the shortness of the toes. Footprints of many of these quadrupeds, particularly those of sauropods like the ones found by Roland T. Bird – and more recently by others such as Dr Martin Lockley of the University of Colorado, Denver – show that their feet left short, round footprints, very similar to those of elephants. That proves beyond doubt that the foot of these creatures was designed around a thick, fibrous wedge heel. Instead of having the long bones of the foot lying flat on the ground so that the heel has to be raised at each step, these bones are artificially raised off the ground by a heel pad.

The advantage of this heel is that it saves an enormous amount of energy which would otherwise have to be used to raise and lower the ankle (therefore lifting the entire body weight up and down!) with each stride. The human foot is plantigrade ("sole-walking"): the sole of the foot, extending back to the ankle, is placed on the ground. Thus every step we take involves raising and lowering the ankle, and as this happens so our whole body moves up and down, wasting muscular energy. We may be able to afford this waste of energy, but a dinosaur weighing 20 tons or more cannot afford to be that extravagant.

It is not surprising to find that the structure of the front legs of quadrupeds is very similar to that at the back, except that – especially in animals with heavy tails, such as dinosaurs – the front legs carry a smaller proportion of the total body weight and are therefore somewhat smaller in overall dimensions and strength.

To appreciate the way in which the backbone of a quadrupedal dinosaur such as *Diplodocus* works, think of a suspension bridge. The front and back legs are the upright towers of the bridge, and the backbone is the roadway running between them. In an actual suspension bridge the "deck" or roadway is supported by cables slung between the towers. Similarly, the backbone of the dinosaur is supported by the ligaments and muscles of the back. But the backbone of a *Diplodocus* has to deal with something that the designer of a bridge never dreamed of, for the creature does not just stand there: it walks. The backbone must be flexible, and its vertebrae are jointed together in a most complicated manner to allow the neck, back and tail to bend.

Between the hips and shoulders, the backbone is bowed upwards slightly; in fact, the deck of a suspension bridge is also slightly arched, and in both cases the arch helps to spread the weight towards the ends. But unlike the deck of a bridge, the backbone is also braced from below by the ribs, which are to some extent supported by muscles running up from them to the shoulders and hips.

The large tail is a typically reptilian feature and provides anchorage for the powerful leg muscles; but in the case of the giant sauropod it has another very important function, in which it works together with the neck. The spines on the vertebrae of the backbone are large in these dinosaurs, and often have a groove running down their middle. The most likely reason for this is to anchor a massive, slightly elastic tendon which ran between the tail and the neck and over the whole length of the backbone between them. This arrangement allowed these dinosaurs to have an economical system for supporting both neck and tail, which did not require vast amounts of muscular power. In life these animals would almost certainly have been able to stand at rest with both neck and tail suspended clear of the ground with very little effort.

The neck of the sauropod could have been raised and lowered by using local muscles within the neck so that these creatures could reach high into the trees to find their food. And, as Bakker suggested, it seems likely that these animals could rear up on their hind legs, using the tail as a "third leg" to make a tripod. Calculations based on the cross-sectional area of the hind legs of these dinosaurs and the known strength of bone, made by the British Professor Robert McNeill Alexander of Leeds University, suggest that the hind legs were certainly strong enough to support the full weight of the body.

In addition to the great complexity of the support system for the backbone in these massive animals, another factor of great importance comes into play: weight saving. Paradoxically, land animals weighing 10, 20 or even 30 tons must try to keep their total body weight to a bare minimum – every small saving reduces the cost of supporting the body. As was seen in some of the larger theropods, vertebrae often have holes in their sides (pleurocoels). In sauropods these become very elaborate structures reducing the backbone to a virtual honeycomb. Even the ribs of some sauropods are hollow in order to save weight.

Using the example of an extremely large, long-necked, and long-tailed sauropod is perhaps a little biased, but these general principles of construction apply to all quadrupedal dinosaurs. Smaller ones are naturally not under such extreme pressure to economize on their use of bone. Some of the Triassic and Early Jurassic sauropodomorphs do not have pleurocoels, and do not show such extreme development of suspensory tendons in the neck and tail. No ornithischian has pleurocoel cavities in its vertebrae, for example; and, as already mentioned, the suspensory ligaments in the backbone are formed by the ossified tendons, rather than the single long tendon of large sauropods. No ornithischians developed long necks; if anything they went the other way, tending towards shorter necks and tails, and larger, heavier heads.

How did they feed?

There is a very obvious division into two types of feeding: carnivorous, meat eating; and herbivorous, eating plants. The greatest diversity of dinosaurs is to be found among the herbivores, which include representatives from both the major groups of dinosaurs, the saurischians and ornithischians. The only carnivores were the theropod saurischians, and although they were relatively conservative in body design they were far from conservative in the ways they went about obtaining their food.

Carnivorous theropods

Most theropods were meat-eaters. These share the same kind of teeth. All are narrow, sharply pointed and curved, with the back edge, and in some cases the front edge as well, marked by fine serrations much like those on a serrated kitchen knife. Such teeth are ideal for penetrating the flesh of their prey to bring about a speedy kill.

Feeding must have been a merciless process once the potential prey had been spotted by the sharp-eyed, fleet-footed predator. A rapid dash, and the sharp claws and teeth would have immobilized the victim. Once the jaws were clamped on the prey the serrated edges of the teeth would be very effective at tearing the flesh, so that

155

Daspletosaurus, *a large tyrannosaur, has failed to catch the large, fearsomely defended ceratopian* Styracosaurus *unawares. With time to avoid the desperate lunge of the predator the prey now has the advantage and may spear the oncoming foe with its large nose horn.*

large, bite-sized pieces could be removed. The struggles of the prey would actually help the predator to bite by forcing the curved teeth deeper into its own flesh as the animal tries to pull itself away. The theropod would also have made backward jerking tugs, using the muscles of its head and neck to help to tear the prey apart. This style of feeding would have been common to a wide range of theropods. Some may, however, have used slightly different techniques.

Extremely small theropods may well have fed upon insects, in which case their teeth would have tended to be smaller and more spiky. Being able to rip into flesh would not have been as important as being able to crack open the tough external skeletons of insects such as beetles and dragonflies; small, spiky teeth are idea for this purpose.

Extremely large theropods, such as the tyrannosaurs, may have fed in a slightly different way from the majority. It seems probable that they may have caught their prey first by stealth, then a final rapid dash. As the top predators of the Late Cretaceous, they probably fed on the numerous herding species of hadrosaur and ceratopian which lived at this time. Both of these prey are large and quite fleet-footed animals, which would have taken considerable effort to subdue; this was especially so for the ceratopians, many of which could defend themselves with fearsome horns on their heads. Thus a surprise attack may well have been almost essential. The massive heads and the extremely long, almost straight teeth of these creatures supports this idea. The prey was probably killed in one deadly rush; the tyrannosaur running headlong at the flank of a large *Triceratops* with its mouth wide open. The impact between the tyrannosaur, weight up to 1 1/2 tons, and a 2-ton *Triceratops* would have been colossal, driving the long, sabre-like teeth deep into the flesh, causing massive wounds and, if the tyrannosaur was on target, rapid death to its victim. Aim and timing would have been vital: a mistimed lunge by the tyrannosaur would have allowed the ceratopian to sidestep the attack and possibly spear the predator on its huge horns.

Once its victim was safely subdued, the tyrannosaur could have consumed it at leisure, using its sharp teeth to slice off sizeable chunks of meat. The jaws of these tremendous creatures were slightly flexible on either side, allowing the jaws to bulge sideways so that large pieces of meat could be swallowed.

The bones in the skulls of most theropods are surprisingly thin and light, enclosing large spaces in the sides of the face. This combination enables the skull to be both light and manoeuvrable as well as very flexible and strong. Large openings at the front for the nostrils and near the middle for the eyes are expected in all skulls, but here other openings are found in front of and behind the eyes which provided areas for the attachment of powerful jaw muscles.

Baryonyx is a theropod which may have fed upon fish, though other animal prey cannot be ruled out.

Some of the larger theropods, such as allosaurs and tyrannosaurs, have skulls which are somewhat more heavily built. The greater weight of bone corresponds to the much larger forces that have to be withstood by the skull bones during prey capture and feeding. There are also hinges which run across the top of the skull just behind the eyes. These hinges allow the snout to tilt up and down slightly. Such a feature certainly allowed some flexibility in the upper jaw while feeding, but it was probably more important as a device to permit a certain amount of "give" to the skull to absorb shocks when the beast ran at perhaps 20 mph (30 km/h) head first straight into its prey.

Baryonyx, discovered in Britain in 1983, is a theropod which seems to break the rules. This 28 ft (8.5 m) long predator has a most puzzling combination of skull characteristics. It has a rosette of very large teeth at the tip of its snout, and behind this is an unusually large number of smaller, almost pencil-shaped teeth. The snout is very long and low, yet the rear of the skull is quite deep. Just what this means in terms of the creature's way of life has been the subject of much debate. The rosette of large teeth at the tip of the snout resembles that seen in large crocodiles, yet the pencil-like teeth are more similar to the teeth of fish-eating creatures. The length and low profile of the front of the skull are also completely unexpected features. It has been suggested that this was a fish-eating (piscivorous) theropod, and that the large, curved claw found with the skeleton may belong to the hand and would have been used as a gaff for spearing fish. Another, slightly more conventional, interpretation put forward was that the animal was a carrion feeder, in which case the long, low snout would have been useful for probing into the body cavity of a rotting carcass.

That is not the only theropod to have caused paleontologists much puzzlement, and will undoubtedly not be the last. New discoveries are constantly taxing our powers of interpretation and imagination. For example, as well as the conventionally toothed theropods there was a variety of toothless types.

The ornithomimosaurs ("bird-imitating lizards") such as *Ornithomimus, Struthiomimus* and *Gallimimus* are so similar in their appearance to living ostriches that it would seem very probable that they lived and fed in similar ways. The beak would have allowed them to feed on practically anything, provided it was small enough to be swallowed: insects; small animals such as lizards, amphibians, and mammals; and perhaps fruits and berries.

The *oviraptorosaurs* ("egg-snatching lizards") such as *Oviraptor* and *Caenagnathus* may well have had feeding habits broadly similar to those of the ornithomimosaurs, but it is possible that they were specialist egg predators. The short, thick, and heavy jaws may well have been adapted specially for cracking open the large, thick-shelled eggs of dinosaurs.

The problems of eating plants

In general, plants are much more difficult to feed on than meat. While meat is readily cut up and easily digested, plant tissues are much tougher and need to be chipped and ground down to release their goodness. What is more, plants are made largely of a material known as cellulose. Unfortunately no animals have ever developed digestive juices (enzymes) capable of breaking down cellulose; this presents herbivores with even greater difficulties should they want to extract the maximum amount of nutritious material from the plants that they eat.

The problem of digesting cellulose has fortunately been solved by microbes which are able to produce enzymes that will turn cellulose into simple sugars which animals can digest. Most plant-eating animals living today take advantage of this fact by hosting colonies of microbes which live quite comfortably in their intestines.

Herbivores, therefore, all tend to show varying adaptations associated with breaking up tough plant food, and providing space in their guts for microbes to live and do their important job.

Sauropodomorphs

The early sauropodomorphs of the Late Triassic and Early Jurassic, known as "prosauropods" and including *Plateosaurus, Lufengosaurus, Massopondylus,* and *Anchisaurus,* were the first tall herbivores ever to appear on land. Before this time, the highest that any plant eater could reach must have been about 4 ft (1.2 m), leaving trees relatively unmolested. The arrival of the first tall, plant-eating dinosaurs must have devastated the trees of the Late Triassic.

The animals themselves had relatively small heads and long necks, and bodies which could be tilted up, so that they could stand on their back legs with ease, giving them a vertical reach of about 20 ft (6 m), or more in some larger varieties.

The jaws were lined with quite small, diamond-shaped teeth with very coarsely roughened edges, which seem well suited to tearing off pieces of vegetation, though they were clearly not specifically adapted for grinding or chewing up this food. Grinding of the plant food seems to have taken place in a special part of the gut, similar to the gizzard of a modern bird. The gizzard was a powerful muscular bag, lined with pebbles known as gastroliths ("stomach stones") which the animal had swallowed deliberately. Once the plants reached the giz-

GASTROLITHS (STOMACH STONES)
Angular polished stones such as these are often found with or nearby the skeletons of the larger sauropodomorph dinosaurs. The stones would have been picked up by the dinosaurs and swallowed so that they could have become lodged in the muscular lining of the gizzard. Such stomach stones would have pounded the toughest of plant food to a fine pulp from which the nutrients could have been extracted. This would have been the first stage in the digestive process, and would have been followed by fermentation by microbes further back in the gut.

zard they were whirled around and pulverized by powerful contractions of the thick muscular walls of the gizzard and the grinding action of the gastroliths. After the plants had been reduced to a thick soup-like liquid they would have passed down to the rest of the intestine where nutrients could be absorbed, and microbes could further digest the cellulose.

Evidence that this actually happened comes from the discovery of piles of angular, polished stones within the rib cages of some early sauropodomorphs such as *Massopondylus*, and showing the approximate position of the muscular stone-lined gizzard. Crocodiles today have a powerful muscular stomach, and they too swallow stones which not only act as ballast but also help to crush up the bones of their prey.

The later Jurassic was the time of appearance of the greatest plant eaters that have ever walked the Earth, the sauropods. These animals took the developments seen in the earlier forms to bizarre extremes. Returning to a stance on all fours, these creatures developed massive bodies to house a huge stomach and intestine for processing vast quantities of vegetation. The neck was extremely long to enable sauropods to reach into high trees, and was counterbalanced by the long tail. The head was notably small.

Brachiosaurus is one of the best known examples of a giant sauropod. Giraffe-like in its general body shape, but immensely larger than any giraffe, it stood some 35 ft (11 m) high on its very long front legs, giving it a marvelous reach into the tree tops. The entire animal probably weighed between 20 and 30 tons and was 75 ft (23 m) long, yet the skull was only about 3 ft (0.9m) long – a large skull perhaps by normal standards, but not for an animal that size!

The jaws are quite strong, and the teeth are roughly spoon-shaped, with chipping edges for biting off leaves and twigs. However, as with the earlier forms, the head was only the food gatherer: its sole purpose was to keep a constant supply of plant to the stomach below. A gigantic muscular gizzard, lined with pebbles, pounded up the plant food until it was ready to be passed back to the rest of the intestine. Here the food would have much of the readily removed nutrients removed, leaving the indigestible cellulose and other plant materials behind. The latter would then pass into a series of large, blind-ending pouches springing from the sides of the main intestinal tube. These pouches, known as caeca, were the chambers which housed the gut microbes whose job was to ferment the cellulose. The food remained in this part of the intestine until most, if not all the nutrients had been removed, before being passed out of the body as a dropping. This highly efficient system ensured that these gigantic animals extracted all the nutrients possible from the low-quality food on which they relied.

Aerial heads
*Perched high on top of the neck, the head cropped
leaves from the tree tops. The teeth were spoon-shaped,
with sharp edges, ideal for cutting, but of little use for
chewing tough types of food.*

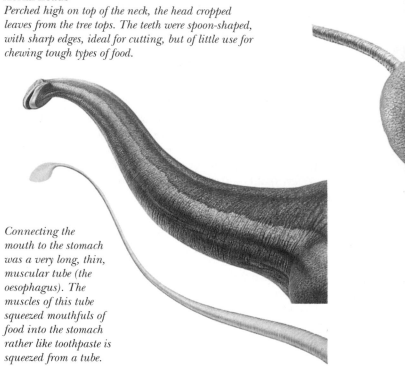

*Connecting the
mouth to the stomach
was a very long, thin,
muscular tube (the
oesophagus). The
muscles of this tube
squeezed mouthfuls of
food into the stomach
rather like toothpaste is
squeezed from a tube.*

Ornithischians

From the outset ornithischians appear to have taken a rather different approach to feeding on plants. All have a sharp horny beak at the
front of the mouth, which would have been very effective at cropping
plants. There was probably some partitioning between these
dinosaurs of the types of plants upon which they fed, based upon the
shape of the beak. Narrow, pointed beaks were clearly suitable for
specialist feeders capable of picking off fruits or tender shoots individually. Ornithischians with broader muzzles probably fed in a more
rough and ready manner, taking mixtures of leaves, twigs, bark, and
anything else caught between the guillotine-like edges of the beak.
Behind the beak the jaws were equipped with varying arrays of teeth,
depending upon the toughness of the food.

The most simple type of ornithischian tooth was approximately diamond-shaped with coarsely serrated edges, and rather similar to those
of some of the early sauropodomorphs. Such teeth, as we have seen,

Dinosaur guts

LEFT: *A diet of tough vegetation required a very powerful digestive system in order to extract the essential nutrients. Roughly cut plant fragments passed straight into the large muscular stomach where the vegetation was mixed with digestive enzymes and pounded through strong squeezing movements and the action of stony gastroliths embedded in folds in the stomach wall.*

BELOW: *A window into the stomach shows the thick muscular lining and the embedded stones (gastroliths) used to reduce the toughest plant food to a thick paste.*

ABOVE: *Once the stomach has reduced the food to a thick paste it can be passed further back along the gut. At first it passes through the narrow intestinal tubes and then into the large sack-shaped caeca. These large, thin-walled tubes house colonies of microbes which spend their time growing and breaking down the tough cellulose plant tissue coming from the intestine.*

are well designed for tearing plants but not for grinding or chewing plant food up in the mouth. As a group, the ornithischians included both dinosaurs which did very little chewing in the mouth and those which developed very powerful grinding teeth and jaw muscles.

One additional and very important characteristic of all ornithischians is that they had fleshy cheeks covering the sides of the mouth. No living reptiles have cheeks – only mammals do today – and it is another unusual and special feature of this type of dinosaur. The presence of cheeks is suggested by the fact that there are deep recesses on the sides of the faces of these dinosaurs along the line of the jaw, and the edges of these recesses or pouches are faintly marked with muscle scars showing the position of the edges of the cheeks. Cheeks prevent food from falling out the sides of the mouth and being wasted.

Small ornithopods such as *Lesothosaurus*, *Thescelosaurus* and *Orodromeus*, and the pachycephalosaurs, stegosaurs, and ankylosaurs, all seem to have had a relatively simple feeding mechanism.

The small ornithopods and the pachycephalosaurs had narrow beaks for very precise feeding, and rows of fairly simple, diamond-shaped teeth in the jaws behind. It seems clear that they would have fed upon soft and relatively nutritious food which did not require a great deal of hard chewing to extract the nutrients. This combination of factors suggests that they did not need great stomachs to ferment quantities of cellulose, because the food was of high quality and soft.

Stegosaurs and ankylosaurs were larger and heavier animals than the previous two groups, and were quadrupedal. Although they did not reach the size of sauropods, nevertheless the main part of their bodies was built to the same kind of design: four pillar-like legs support the arched back, from which are suspended the chest and capacious belly. The muzzle is rather broader than in ornithopods, especially in the case of ankylosaurs, which seem to have broad, shovel-like beaks – they clearly consumed a much wider variety of foliage. The teeth are relatively small and little different from those of small ornithopods, and thus seem not to have been used for chewing, simply for cutting up food ready for swallowing.

There is no strong evidence for gastroliths (stomach stones) in any of these creatures. Both of these groups of dinosaurs seem among the least agile and least energetic of all, and it may be that they simply left

ORNITHOPOD TEETH AND JAWS – *IGUANODON*

The skull of Iguanodon *(right) is long and somewhat horse-shaped. Behind the horny beak are rows of sharp cutting teeth, covered over by a muscular cheek. The teeth meet at a steep angle (below – skull drawn in cross section) forcing the upper jaws outwards as the teeth were pressed together, creating a grinding motion which pulverized the food.*

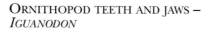

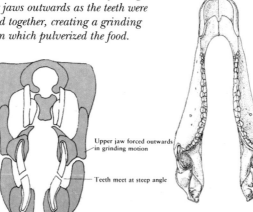

Upper jaw forced outwards in grinding motion

Teeth meet at steep angle

ABOVE: *An upper grinding tooth of* Iguanodon.

LEFT: *A view from above of the lower jaw and teeth.*

the food in their stomachs for longer, providing plenty of time for digestion and microbial fermentation to take its natural course, without needing the assistance of a stone-laden gizzard. The stomach would then need to be large simply so that it could act as a fermentation tank for all the food these big creatures ate over a considerable time.

Two groups of ornithischians deserve particular attention: the larger ornithopods and ceratopians. Both groups had perfected methods of pulping food in the mouth in ways that are reminiscent of those used by some of the most proficient chewers of tough vegetation today: horses and cattle.

Larger ornithopods include creatures such as *Iguanodon, Ouranosaurus,* and the hadrosaurs. The beak is sharp and broad, but behind this region the teeth, instead of sitting in simple lines along the jaw, are arranged into interlocking batteries which wear down to produce a long, pavement-like grinding surface.

During chewing, the lower jaws of these animals swung up and down and, as the teeth engaged and began to bite on the food trapped between them, the upper jaws, which were hinged, were pushed outwards. This curious mechanism of the bones of the side of the face has been discovered only very recently and was first noticed in *Iguanodon.* It allowed the upper teeth to slide sideways across the lower teeth in a grinding motion, which is essential for grinding up tough plants. A similar effect is achieved by horses and cattle (and indeed by ourselves) by moving the lower jaw from side to side, rather than the upper jaw.

The combination of special methods of grinding the teeth together, powerful jaw muscles, and cheeks allowed these dinosaurs to pulp the toughest plant food in the mouth, so that it could be passed back to the stomach for much faster and more direct digestion and absorption. A stone-laden gizzard would not have been necessary, but there would still have been fermentative cacca behind the stomach for microbial digestion of the cellulose.

The **larger ceratopians** had skulls constructed very differently from those of ornithopods. The bones were rigidly held together, with no possibility of movement or hingeing of the upper jaws. The rigidity of the skull was absolutely essential because it needed to support the massive crest at the back, and in many cases huge horn cores over the eyes and on the nose. As far as feeding was concerned the jaws were very distinctive. The beak was extremely narrow, consisting of sharply hooked upper and lower parts arranged and looking rather like those of a parrot. This type of beak was clearly designed specifically for cutting, rather than for cropping or browsing as are the vast majority of ornithischian beaks.

CERATOPIAN TEETH AND JAWS – *TRICERATOPS*

Triceratops *has an extremely heavily built skull (right) with a large, hooked horny beak for nipping off tough plants. The sides of the jaws were covered by muscular cheeks, inside which were the sharp cutting blades made up of hundreds of interlocking teeth* (BELOW – *view looking down on lower jaw). The jaws were operated by massive muscles on the sides of the skull in front of the frill.*

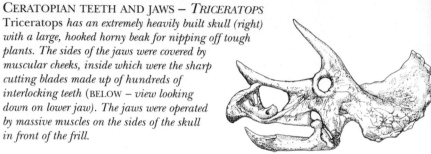

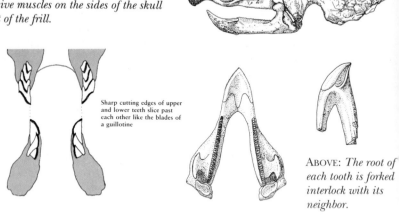

Sharp cutting edges of upper and lower teeth slice past each other like the blades of a guillotine

ABOVE: *The root of each tooth is forked interlock with its neighbor.*

Behind the narrow beak the jaws were lined with hundreds of teeth. Again these teeth form interlocking sets of batteries as seen in the larger ornithopods, but instead of forming pavements of teeth which acted like millstones, the teeth of the ceratopians are very sharp and locked together in such a way that they form guillotine-like blades. They must have been used to slice the plants into short sections by being passed alternately in and out past the teeth – a large, muscular tongue forcing the food out past the teeth into the cheek, then the muscular cheeks squeezing it back into the centre of the mouth.

When the plant food had been sliced sufficiently finely it would have been swallowed and processed in the stomach. It seems likely that these ceratopians may have also used a gizzard to pulp the sliced food more thoroughly before digestion. Although no firm evidence exists for gastroliths inside skeletons of large ceratopians, there is a celebrated example of a fine skeleton of the small ceratopian *Psittacosaurus* found by the American Central Asiatic Expedition to Mongolia in the 1920s; this has a large pile of more than 100 small pebbles in just the area where a stomach might be expected. Large fermentation sacs will also have been present farther back along the intestine, and could certainly have been readily contained within the very large abdomen of these heavily built dinosaurs.

The plants fight back

Herbivorous dinosaurs were clearly devastating "predators" of plants. The armaments they would have used for stripping plants of their foliage were many and varied: horny beaks of different shapes and sizes, chisel-shaped teeth, comb-like teeth for literally raking in leaves or pine needles. Not surprisingly the plants evolved to meet this attack by putting up their own defenses: making themselves spiky, extremely tough and difficult to eat or, in some cases, by developing chemical defenses – substances which can make plants unpalatable, or can actually poison any animals that try to feed upon them. Some of these defense mechanisms proved extremely successful – indeed so successful that they can be seen to this day.

Ferns of various sorts are abundant today, as they were in the Mesozoic, but there are few animals capable of feeding upon them now. Although they are relatively soft plants, their chemical defenses make them extremely unpalatable to herbivores. Yet these were most likely the staple diet of many low-browsing dinosaurs during the Mesozoic. Cycads, which were abundant during the times of the dinosaurs, still live today and are notable for their tough and abrasive leaves. Monkey-puzzle trees (Araucaria) are another example of an ancient type of tree which was common during the time of the dinosaurs. It too is characterized by tough, spiky foliage. Conifers (the pines, firs, spruces, yews, larches, sequoias, cypresses) are also an ancient group of dinosaur times. The narrow, leathery foliage of these plants and the waxes and other chemicals they produce make them more or less inedible.

HOW THE PLANTS SURVIVED
Some of the plants that co-existed with the dinosaurs had defense mechanisms which protected them from the ravages of the herbivorous dinosaur, ensuring their survival even to this day. The protective strategies they developed included mechanical defenses (cycads, below left), chemical defenses (horsetails, centre), and some just grew more quickly (magnolias, right).

167

In all these instances there are few if any herbivores today which feed upon these plants in a comprehensive way. Various rodents (especially squirrels) may feed upon the cones of pine trees, as do birds such as crossbills; both are capable of extracting the nutritious seeds from mature cones. But these are all specialist feeders and not at all comparable to large-scale browsers such as the giraffe, antelope, elephant, or rhinoceros which live on the African plains today. These remarks do not, of course, apply to insects or smaller organisms. These reproduce rapidly and can evolve much faster than larger creatures. Whatever the plant, it is almost certain that some small creature will have developed a strategy for feeding on it.

It is tempting to suggest that what we see today in these plants is the remnants of defense strategies developed against herbivorous dinosaurs. We have no real way of knowing how successful such defenses may have been against dinosaurs. However, since the extinction of the dinosaurs no other group of herbivores seem to have developed a way of successfully feeding upon these plants. They seem to be literally survivors from the time of the dinosaurs, and provide us with a small window on their world.

DINOSAUR BLOOD

Blood, obviously, does not fossilize. But surprisingly, quite a lot can be discovered about how it moved about the bodies of dinosaurs, and how it was used. This has an important bearing on considerations of the physiology of dinosaurs – in particular the controversy over whether they were "warm-" or "cold-blooded."

A dinosaur's heart

In 1980 it was demonstrated that the posture of dinosaurs made special demands on their blood supply. The most important factor in this is the position of the head. To explain this it is probably best to take the most extreme example from the dinosaur world – but the same principles apply to nearly all dinosaurs in some measure. The example I shall use is *Brachiosaurus*, the giraffe-like sauropod.

Brachiosaurus stands over 35 ft (11 m) tall. The brain which controls the activities of this monstrous animal is approximately 25 ft (7.5 m) higher than the heart. In order to pump blood up to the brain at this level, the heart must have had to generate an enormous pressure, and so it must have been huge and highly muscular. While this is all perfectly reasonable and understandable, the fact that the heart was capable of producing enormous blood pressures does not fit comfortably with our knowledge of the way the heart works in a modern reptile.

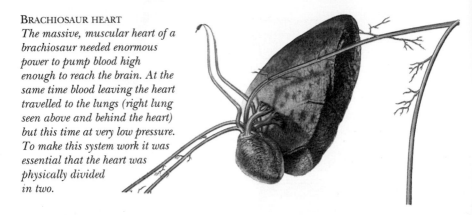

Reptile, mammal, and bird hearts all circulate the blood around the body of the animal in two separate circuits. One circuit passes through the lungs, so that the blood can dump waste carbon dioxide, and take up fresh oxygen. The other circuit takes the freshly oxygenated blood from the lungs right around the body. In mammals and birds these two circuits – the pulmonary circuit (to and from the lungs) and the systemic circuit (to and from the rest of the body) – are physically separated in the heart, which is divided into left and right sides. Living reptiles have hearts which are only imperfectly divided. They have, for various reasons, a "leaky" system which allows blood to leak between the pulmonary and systemic systems within the heart.

If dinosaurs had had hearts similar to modern reptiles, the result would have been catastrophic! The reason for this is apparent from the example of *Brachiosaurus*. Blood at incredibly high pressure in the systemic circuit destined for the brain and the rest of the body would be able to leak into the pulmonary circuit destined for the lungs. The lungs, however, must work at low blood pressures because the gas exchange can only occur effectively across very thin, and therefore very weak capillaries. High-pressure blood would cause massive bleeding into the lungs, and the brachiosaur would have literally drowned in its own blood!

This simple fact implies that most dinosaurs – certainly any that were fairly large and adopted a posture in which the head was held significantly higher than the heart – must have had a fully divided heart similar to that found in mammals and birds. If this reminds you of some prophetic words by Richard Owen in 1841 (page 74) then I congratulate you on the attention you have been paying to this book.

This conclusion may also remind you of the claims of Robert Bakker concerning the mammal or bird-like physiology of dinosaurs. However, as I shall argue later, it is not necessary to assume that dinosaurs were necessarily endothermic simply because they had a fully divided heart.

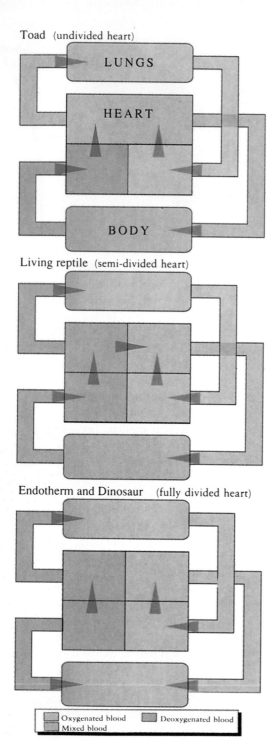

Toad (undivided heart)

LUNGS

HEART

BODY

Living reptile (semi-divided heart)

Endotherm and Dinosaur (fully divided heart)

Oxygenated blood   Deoxygenated blood
Mixed blood

BRACHIOSAUR RESTORED.
This dinosaur (above) had a very giraffe-like posture, with the head craned high over the chest and pillar-like legs. This skeleton anatomy imposes a number of limitations on the way in which the bony anatomy, respiration and blood circulation could have worked. The height of the head required a very powerful heart to generate sufficient pressure to circulate the blood to the brain. The advantage of this is that a high-pressure blood system would have allowed for a far more active lifestyle than could have been expected if this animal had been operating as a traditional reptile.

LEFT: *The toad heart is the simplest, with no division of the main pumping chamber, living reptiles have a partly divided chamber, while living birds, mammals (and probably dinosaurs) are fully divided. This enables them to pump blood much more efficiently and allows them to pursue more active lives than reptiles.*

170

Heating and cooling

Apart from the obvious functions of blood, which are concerned with moving nutrients and gases around the body, another equally important function is that of transferring heat around the body. Evidence for such a use has come to light in recent work on dinosaurs. Dr James Farlow of the University of Indiana, Fort Wayne, and colleagues have looked in detail at the structure of the large plates running down the back of *Stegosaurus*.

There has been considerable controversy concerning the positioning and function of these plates. They seemed to be protective, so it became popular to arrange them either in a tall row like a Red Indian chief's head dress (as they appear in all museum skeletons) or sticking out sideways, partly covering the vulnerable flanks. There was also much debate about whether they should be placed in a single row, in pairs to form two rows, or in two rows of alternating plates. Farlow and colleagues challenged all these views and suggested a completely different use for the plates.

Examination of the sides of the plates, their bases, and even cross-sections of individual plates proved beyond doubt that these bones

The large plates on the back of stegosaurs may have served as signalling devices, but more importantly they acted as solar panels or radiators to control body heat in these dinosaurs.

171

were not solid protective plates at all; they were light honeycomb structures. They seemed to be designed to allow enormous quantities of blood to pour through the plates and out onto the surface of the plates beneath the skin. Why? Farlow suggested that this was to allow the plates to be used either as solar panels to absorb heat, or as radiators to lose excess heat to the air. He was able to back up this proposal by wind tunnel experiments, which also had a bearing on the probable arrangement of the plates. The tests showed that the optimal design for the plates if they were used as radiators was approximately a diamond shape – as indeed they were. It also showed that the best arrangement of the plates for heat loss and heat uptake was in a a staggered series of two rows.

The remaining problem of interpretation for stegosaurs revolved around the endothermy–ectothermy debate. Both factions claimed stegosaur plates for their own arguments. They were either seen as heat radiators to prevent endothermic dinosaurs from overheating, or simply as a sign that these dinosaurs operated like gigantic ectothermic lizards, using their plates to speed up heating of the body in the sun. In fact the evidence balances, so it cannot really be used effectively by either side.

Necks and tails

Another argument concerning heat gain and loss in dinosaurs has concentrated simply on the body proportions of dinosaurs. Sauropods, for example, are rather like elephants in terms of their body proportions if you ignore the neck and tail. Elephants, being bulky creatures, cannot lose heat very quickly through their skin, so if they are very active and begin to overheat they flap their ears vigorously to radiate blood heat very quickly – this is one of the main reasons why elephant ears are so big. Sauropods did not grow big external ears, but they did have long cylindrical necks and tails which provide a very large surface area for heat exchange, and it is possible that they used these to offload heat following strenuous activity.

Bone structure

It is possible to see the passages for blood vessels in specially prepared thin sections of dinosaur bone. Such sections can also be used to draw implications about the physiology of the creature. Some bones, when sectioned and examined under the microscope, show quite dense bony tissue with relatively rare passages for blood vessels; other bones can show very numerous channels for blood vessels. The difference in number of blood vessels correlates very roughly with the "activity" of the bone – that is, how quickly it was being made by the body – and this also reflects the level of activity of the animal. A more active endotherm, for

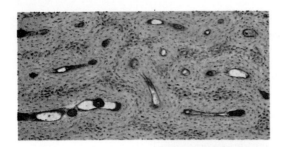

TOP: *Fibro-lamellar bone (without growth rings) was formed by many dinosaurs. It is now seen in large fast-growing mammals and birds.*

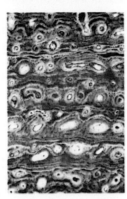

LEFT: *Zonal bones (with growth rings) formed by some dinosaurs during growth. It is now seen in ectothermic "cold-blooded" reptiles.*

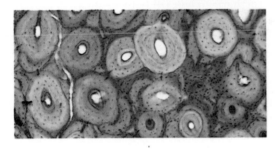

BOTTOM: *Haversian bone, produced by internal reconstruction and often extensive in dinosaurs, is now mainly in large mammals.*

example, would tend to have more blood vessels in its bone than an ectotherm (in scientific terms, the bone would be more vascular).

At its crudest, the argument about the internal bone structure of dinosaurs was used by Bakker to support the idea that dinosaurs were "warm-blooded." As explained earlier, sections of dinosaur bone seem to resemble modern mammal bone more than they resemble modern reptile bone, so he drew the conclusion that dinosaurs had mammalian physiology – or, in other words, were endothermic. Such categorical statements provoked an enormous amount of painstaking work on dinosaur bone structure by a number of paleontologists, most notably the French Professor Armand de Ricqlés at the University of Paris; and, in Northern Ireland, Dr Robin Reid at The Queen's University, Belfast.

At first Professor de Ricqlés strongly favored the view that the bone structure of dinosaurs showed that they had an endothermic physiology. However, continued study of the bones of dinosaurs and of living animals has revealed that the situation is not nearly as clear-cut as was first thought. For example, highly vascular bone can be found in living ectothermic reptiles, and very poorly vascularized bone can be found in small mammals and birds, which have very high activity levels indeed – exactly contradicting the original conclusions! It now appears that highly vascular dense bone can be of at least two types.

The first of these is "primary bone," the material which is first formed as a bone begins to grow. Bone is laid down within a dense network of blood vessels which carry in the materials such as calcium phosphate that form the bone itself. Viewed under the microscope, such bone naturally appears highly vascular. "Secondary bone" gradually replaces primary bone layers from within. This type of bone is formed as channels are opened through the layers of primary bone and new thread-like ribbons of secondary bone are laid down in its place – the channels are known as Haversian canals. These threads of secondary bone are specially designed to withstand stress, and therefore seem to strengthen the bone as it grows and has to support greater weight and muscular forces exerted by the body.

Dinosaurs, it would seem, have all types of bony tissue when looked at in detail, and this can vary enormously even within a single bone in the body, or between bones, depending upon the role which they play in the life of the animal. Dr Reid has shown that dinosaurs are notable for the amount of primary bone tissue in their legs, which suggests that they grew very quickly when they were young. In later life this is sometimes found to have been replaced by secondary bone. This process is much like that in mammals and birds, and supports the idea that dinosaurs were highly active, fast-growing creatures. However, he has also found seasonal growth rings in some bones, rather like the growth rings that can be counted in the cut section of a tree trunk. This suggests that some dinosaurs tended to be more active during parts of the year, thereby growing more quickly; and less active at other times, growing more slowly and therefore depositing bone more slowly, thus producing the alternating lines of growth rings. This pattern of growth is much more like that expected in an ectotherm whose activity levels are governed by seasonal changes. This contradictory evidence has been used as evidence that dinosaurs, far from having either a typical reptilian ectothermic physiology, or a mammal- or bird-like physiology, had a metabolism unique to themselves.

A further complication has emerged very recently from some work done in America by Professor de Ricqlés on a growth series of dinosaurs collected by Dr John Horner of the University of Montana, Bozeman, from sites in western Montana. This work has revealed that some dinosaurs evidently grew extremely quickly when they were young, reaching adult size in a surprisingly short number of seasons. However, once they had reached full size, it seems that their metabolic rate (as far as growth at least was concerned) rapidly slowed down. Thus it could be argued that dinosaurs grew like birds or mammals, very quickly, but when they reached full size their activity levels dropped dramatically so that they came more to resemble living reptiles. Robin Reid adds further information to the debate by pointing out that some of the dinosaur bones which he has sectioned suggest that some other dinosaurs continued to grow actively throughout their lives, with no indication of any slowing when they reached adulthood.

We must be cautious about jumping to conclusions based on first observations – Nature is almost always more complicated than we imagine. It seems that dinosaurs were certainly capable of high rates of growth, which correspond to those seen in mammals and birds today. There is also some evidence that dinosaur activity levels were not necessarily always high, a sign that they may have been endothermic. Perhaps dinosaur physiology was "dinosaurian" as Dr Reid suggested, rather than exactly like that of a mammal, bird or reptile.

INTELLIGENT DINOSAURS?

Dinosaurs' stupidity is a byword – but is that view well founded? The question was investigated by Professor James Hopson of the University of Chicago in the 1970s. A large number of skulls of dinosaurs have been preserved well enough for us to make estimates of the volume of the brain case – the bony box in which the brain sits within the skull. Allowing for the fact that the brain does not necessarily fill the entire brain cavity, it is possible to compute the approximate volume of the brain in a number of dinosaurs. The answers which emerged were important because the brain and its development control the activity and social life of the animal. Furthermore, a large brain needs a constant body temperature to keep it working properly.

The results were interesting and unexpected. Brain size does not have to increase directly in proportion with body size, and the giant sauropods with their relatively small brains did not turn out to be as brainless as was thought; they had brains about as large as would be expected in a reptile that size. More interestingly, some of the smaller, more agile theropods and ornithopods had exceptionally large brains, more comparable in proportion to body size with those of modern birds and small mammals. Some of the smaller and less agile types, such as stegosaurs and ankylosaurs, seemed to have relatively small brains for their size.

On average dinosaurs proved not to be as stupid as they had been made out to be from the statements of people such as Marsh (see page 96). If anything, they had brains comparable to those of modern reptiles, and a few of the more active ones seem possibly to have had an intelligence more comparable to that of mammals and birds. Evidently the more active a dinosaur appears from its anatomy, the more likely it is to have a large brain.

HOW DINOSAURS BEHAVED

Brain size and bodily control bring us to one of the most interesting of subjects concerning dinosaurs: their behavior and social life. The past twenty years have seen an enormous growth in clues which can allow us to piece together aspects of dinosaur behavior. Some of these have been astonishing.

The family life of ceratopians

The first real insight into the social life of dinosaurs came from Mongolia with the discovery of *Protoceratops* and its nest sites in the 1920s. The eggs and nests showed that these dinosaurs took considerable care to arrange their eggs in neat concentric rings before they buried them in mounds. In addition the presence of many dinosaurs near these nest sites indicated that there was considerable social activity

ABOVE: *Discoveries of nests of eggs of the dinosaur* Protoceratops *revealed that they were laid in approximately concentric circles. Upwards of 30 eggs have been found in some nests, which suggests that females may have laid their eggs in communal nests, indicating cooperative behaviour.*

A VARIETY OF CERATOPIAN SKULLS

Centrosaurus
RIGHT: *The large, rhinoceros-like horn of this dinosaur was undoubtedly a superb defensive weapon but within a herd it was used as a signalling device.*

Pachyrhinosaurus
LEFT: *The roof of the snout has a large thickened boss of bone instead of a sharp horn.*

around the nests. Also, the large range of skeletons made it possible to draw tentative conclusions about the manner of growth of these dinosaurs – particularly by charting the changing proportions of the skull and crest as size increased – and also allowed attempts to identify the sex of at least some of the creatures. It seemed unlikely that dinosaurs living in the same place at the same time, but looking slightly different in terms of the height of the snout or size of the head frill, could be anything other than males and females of the same species.

Follow-up work on *Protoceratops* by Dr Peter Dodson of the University of Pennsylvania used statistical analysis to confirm the proposed sexual differences between the different skull types and the growth pattern displayed by the skulls. This was followed by a broader survey of ceratopians in collaboration with Dr James Farlow of the University of Indiana, Fort Wayne. They were looking in particular at the function of the frill and horns in these animals – the features which give them such a distinctive appearance.

For a long time it had been though that the horns and frills of ceratopians were used simply as defensive weapons and protection. The discovery of *Tyrannosaurus* in rocks of the same age as *Triceratops* seemed strong evidence of the need for protection of the neck by the long frill and large defensive horns. The real problem, however, was to explain the remarkable diversity of frill and horn types seen in ceratopians. Professor Edwin Colbert of Flagstaff, Arizona suggested that the diversity was merely random, each ceratopian developing its own pattern of horns and frills to protect it against predators, much as in the varied horns of African antelopes.

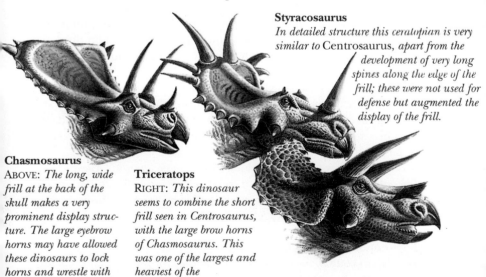

Styracosaurus
In detailed structure this ceratopian is very similar to Centrosaurus, *apart from the development of very long spines along the edge of the frill; these were not used for defense but augmented the display of the frill.*

Chasmosaurus
ABOVE: *The long, wide frill at the back of the skull makes a very prominent display structure. The large eyebrow horns may have allowed these dinosaurs to lock horns and wrestle with each other.*

Triceratops
RIGHT: *This dinosaur seems to combine the short frill seen in Centrosaurus, with the large brow horns of Chasmosaurus. This was one of the largest and heaviest of the ceratopians.*

177

A totally different explanation was put forward by Professor Ostrom at Yale, when he suggested that the frill behind the skull served as an anchor point for immense jaw muscles. Unfortunately, although there may be some truth in this proposal, it does not explain why such frills should be so variable in shape – a common muscular function would tend to lead to uniformity of shape – and why some are adorned with long, delicate spikes, as in *Styracosaurus*, which clearly have nothing to do with jaw muscles.

Dodson and Farlow suggested in their review of horns and frills that ceratopian headgear was variable because it was important for the social life of these animals. Work on modern antelopes had suggested that horns, although useful for defence, were important mainly to behavior. The size and shape provided signals for recognition between individuals in herds, and established a "pecking order" for the maintenance of territories, especially by males during the rutting season. In the larger-horned species, the horns could be used in head-to-head pushing contests of strength, or simply as "flags" which could, for example, be brandished at rivals. Dodson and Farlow proposed that this was also the case for ceratopians, a view widely accepted today.

Some ceratopians such as *Protoceratops* with blunt nose horns and short crests probably interacted by waving their short frills in the air, by lowering the chin and swinging the head from side to side. If that failed, they may have resorted to delivering sideways blows with the head to the flanks of opponents. Larger but short-frilled ceratopians such as *Centrosaurus*, *Monoclonius*, and *Styracosaurus* tended to have very large single nose horns rather like those of a modern rhinoceros. Such horns would have been lethal if used for fighting. To avoid such deadly contests these animals probably relied on displays, bluff, and evasive manoeuvring. Such visual clues must have been important because there is circumstantial evidence that these animals lived in immense herds during the Late Cretaceous. The theory is also supported by the frill of *Styracosaurus*, which has long, delicate horns projecting from its edge which would have been easily damaged in a serious fight and so must have been used mainly as visual signals.

Long-frilled ceratopians such as *Chasmosaurus*, *Anchisaurus*, *Torosaurus*, and *Triceratops* had relatively long frills and a much reduced nose horn, but extremely well developed eyebrow horns. Tucking the chin in and waggling the head would have created a very impressive display, greatly reducing the need for combat. However, in the event of a standoff between two individuals it is possible that the brow horns could have been locked together for pushing and twisting in a contest of strength.

The crests of hadrosaurs

Many of the theories put forward to explain the function of hadrosaur crests have been discussed already (see pages 104, 121), but most have been found wanting in one way or another. In the mid-1970s Professor James Hopson of the University of Chicago looked at them again. He argued that the crests were essentially signalling devices linked to the social life and interactions of hadrosaurs, rather than being snorkels or air reservoirs. In an attempt to prove that this theory was correct he drew up a series of logical conditions which had to be fulfilled if he was correct. First, these dinosaurs had to possess acute vision and hearing. Professor Ostrom had already proved this to be true. Second, it was important that the shape of the crest did not correspond to the cavities inside – that is to say, the external shape was more important than the internal structure. This is so in several cases, where the air passages bear no relationship to the crest shape, as in *Lambeosaurus* and *Corythosaurus*. Third, the crest shape should be very specific to each species, and there should be a difference between the crests of male and females. This had been confirmed by the work of Dodson, who had analyzed the proportions of hadrosaurs and pointed out that a number of remains were divided into male–female pairs. Fourth, it was essential that if several species coexisted that they should show absolutely clear differences in crest shape, to ensure there was never any risk of confusion between the sexes of different species. Finally, he predicted that the crest shape should become more exaggerated as hadrosaurs evolved. This prediction does not seem to have proved correct, however, because some of the latest types were crestless.

Despite incomplete fulfillment of his theory, Hopson seems to have been proved largely correct. His idea was further elaborated upon by the idea that the crest may also have had additional uses in the animals' behavior. Some with a crude bump on the nose, such as kritosaurs, were thought to have used this as a simple weapon during male contests of strength. In those with more elaborate crests it seemed likely that behavior was more ritualized to avoid damage to these complicated structures. These would have included simple head-bobbing displays, such as are very popular today with living reptiles; and vocalizing, using the crests as sound resonators in order to generate a loud bellow.

Proof that bellowing was indeed possible soon emerged from the work of Dr David Weishampel at the Johns Hopkins University, Baltimore, who has built a model of the crest of the hadrosaur *Parasaurolophus* and used it to generate the sound that this dinosaur may have been able to produce 75 million years ago. There is a precisely constructed arrangement of passages inside the crest, which give a very deep,

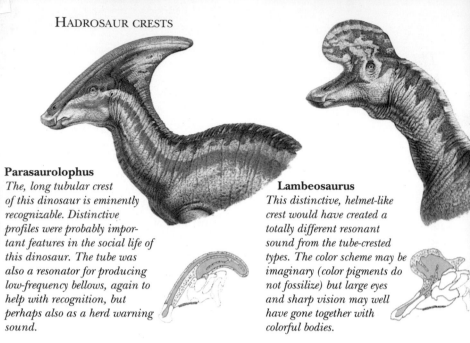

Parasaurolophus
The, long tubular crest of this dinosaur is eminently recognizable. Distinctive profiles were probably important features in the social life of this dinosaur. The tube was also a resonator for producing low-frequency bellows, again to help with recognition, but perhaps also as a herd warning sound.

Lambeosaurus
This distinctive, helmet-like crest would have created a totally different resonant sound from the tube-crested types. The color scheme may be imaginary (color pigments do not fossilize) but large eyes and sharp vision may well have gone together with colorful bodies.

reverberating note. The depth of the note itself may be important for a number of reasons. The note was probably distinctive so that it could be readily recognized by others of the species, and in this respect the timbre may well have been important in establishing some form of "pecking order" within the herd. Another purpose of the low-frequency note is that it could have been used as a warning against predators, alerting the others in the herd. A note of low frequency will travel great distances, and it is also very difficult to identify where the noise came from. That is very important because the last thing that the animal which gave the warning wants is to become the center of attention of the predators.

Other hadrosaurs have more obscurely shaped crests – for example *Lambeosaurus* and *Corythosaurus* – and their sound is less easy to establish. Some, indeed, lack a crest altogether, such as *Anatotitan*, *Anatosaurus* and *Saurolophus*. However, these forms have broad shallow depressions on the sides of the face which may well have formed the floor to large fleshy nostrils that acted as resonating sacs, similar to the inflatable noses of elephant seals.

Hadrosaur nests

During 1978 a remarkable dinosaur find was made, not in the field during an expedition but in a rock shop at Bynum in Montana, by two paleontologists, Dr Robert Makela and John Horner. They found the bones of baby hadrosaurs which had been excavated nearby. The find started a sequence of events that was to attract worldwide attention to the rock formations in this part of Montana.

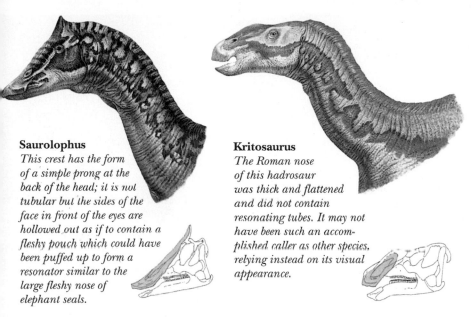

Saurolophus
This crest has the form of a simple prong at the back of the head; it is not tubular but the sides of the face in front of the eyes are hollowed out as if to contain a fleshy pouch which could have been puffed up to form a resonator similar to the large fleshy nose of elephant seals.

Kritosaurus
The Roman nose of this hadrosaur was thick and flattened and did not contain resonating tubes. It may not have been such an accomplished caller as other species, relying instead on its visual appearance.

Makela and Horner traced the baby dinosaur bones back to their original location in the Late Cretaceous rocks of a range of low hills southwest of Bynum. Investigation of the site soon resulted in the discovery of not just baby dinosaur bones, but egg shell fragments, and finally a complete nest containing the jumbled remains of baby hadrosaurs and crushed egg shell. The nest was a shallow, basin-like depression which had clearly been scooped out in the ground, and there seemed to be some trace of vegetation associated with, or perhaps lining the nest. This remarkable discovery set off fieldwork in the area which continues unabated to the present day. It was the first discovery of a nest with babies still inside, and was to lead to surprising insights into the social life of hadrosaurs and the other creatures which lived in this area of the world during the Late Cretaceous.

In the following year, 1979, a proper field crew travelled out to Montana to collect from the same area. They found more nest sites of hadrosaurs, and in a nearby rocky hill – later to become known as "Egg Mountain" – large numbers of eggs of a different ornithopod dinosaur – a hypsilophodontid later named *Orodromeus makeli*. As if that was not enough, the field crews also discovered a bone bed – a layer of rock on which were preserved the scattered remains of a huge herd of hadrosaurs numbering perhaps tens of thousands. Further sites have been discovered since which have made this area one of the world's richest for the remains of dinosaurs and eggs.

The hadrosaur which had laid the eggs at the sites in Montana was new to science and was named *Maiasaura*, for an old Roman mother

HADROSAUR PARENTING *This image of a parent hadrosaur caring for its young around the carefully constructed nest site is one that was unthinkable to most paleontologists as little as 12 years ago. The discovery of nesting colonies, egg fragments and evidence that hadrosaurs reared their young in the nest provide new insight into dinosaur behavior and ways of life, dramatically changing the received views about dinosaurs.*

goddess also known as the Bona Dea or "good goddess." The name was chosen deliberately because of what they found when they examined the nests. These were clearly carefully made, having been scooped out of the earth to make a large bowl-shaped depression. This was apparently lined with soft vegetation into which the eggs were laid. The nests were notable for two features. First, the egg shells were broken into small pieces, which suggested that they had been trampled by the hatchlings. Second, the baby hadrosaurs had worn teeth, which suggested that the young hatchlings had either been fed while in the nest, or had made forays out of the nest and then returned later. This is strong evidence that the parents looked after their young, which had never before been recognized in dinosaurs.

The final proof, if any were needed, that this was the likely nesting behavior of these dinosaurs comes from the nests themselves. The young had in some cases perished in the nest. How otherwise could the presence of their small bones be explained? It became clear that the instinct for the young to stay in the nest was extremely powerful. Perhaps in this case the parent had been killed while searching for fresh food for its young, leaving its young abandoned to die of starvation as they waited patiently for the parent to return.

As work continued at the Montana sites it gradually became apparent that many of the nests which were found together were actually at the same geological level – in fact, they had been made at the same time. This brought with it the further realization that the crews had discovered not just a few odd nests, but a complete dinosaur nesting colony. Dinosaurs seemed to have returned to this area from season to season to lay their eggs in much the same way as birds return to nest sites year after year.

Add to this the discovery of the bone bed indicating the presence of huge herds of hadrosaurs in the vicinity, and it becomes tempting to believe that these dinosaurs migrated in huge herds across North America during the Late Cretaceous. It seems quite possible that vast herds would have followed rich pastures through the year, returning each season to their nesting grounds to rest, lay their eggs, and rear their young. Herding, nesting in colonies, and parental care of the young imply considerable social interaction between these dinosaurs.

But the story does not end here. Associated with the hadrosaur nests and eggs were the nest and eggs of other dinosaurs with which the hadrosaurs shared the area. The hypsilophodontid *Orodromeus* was also found to have laid eggs in nests. The eggs were found in "Egg Mountain" and a little further away on "Egg Island." They proved just as interesting as those of the hadrosaur finds.

Some of the *Orodromeus* eggs and nests were extremely well preserved. Much less trampling of the eggs seemed to have occurred – which was a clue in itself. One nest discovered by Horner and his team was complete and consisted of nineteen eggs laid in a precise spiral, with a single egg placed vertically in the centre of the clutch and the others leaning progressively farther outward along the spiral. Each of the eggs was little distorted and very well preserved. So was there anything inside? The answer was produced by looking at the eggs with a CAT scanner – the machine used for making detailed X-rays of the human head. The scan showed embryonic bones. The researchers painstakingly cut open one egg and separated out the contents. Yet another first for the Montana crew – the first ever dinosaur embryo within an egg!

Comparison of the eggs and nest arrangement of the hadrosaur and hypsilophondontid shed more light on the varieties of parental care in dinosaurs. Many hypsilophodontid nests proved to contain untrampled but opened eggs. This suggested that the young after they emerged from the eggs did not stay in the nest – thereby trampling the egg shells – but left promptly. The implication was that these dinosaurs indulged in little if any parental care of their young. As in the case of many modern reptiles, the eggs were laid in the nest and then abandoned. When the young emerged they were immediately mobile and able to fend for themselves. But was there any supporting evidence? The answer was found in the bone structure of the embryonic dinosaurs.

The joints between the bones of the legs of embryo *Orodromeus*, when examined under the microscope, proved to be well formed and ready for action, which is in accord with the idea that they left the

MONTANA'S DINOSAUR EGGS AND EMBRYOS

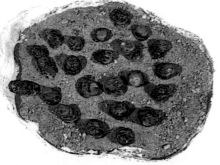

ABOVE: *"Egg Mountain" has revealed entire clutches of Orodromeus eggs, many virtually undistorted.*
RIGHT: *CAT-scan analysis has shown tiny embryo bones inside the eggs. The embryo would have been folded neatly as in this model.*

ABOVE: *This model attempts to recreate the "birth" of a maiasau embryo from an egg. Embryos of this dinosaur have not been found within eggs since unhatched eggs tend to be trampled by hatchlings during their extensive time in the nest.*

184

nest right after hatching. By contrast the maiasaur embryos showed extremely poorly developed joints in their legs, therefore, they would have been nestbound after hatching and fed by the parents.

The theropod *Troodon* was also present in this area. Teeth and jaws and partial skeletal remains of this fleet-footed dinosaur have also been found associated with or near the nesting sites of both ornithopods. This was undoubtedly an opportunistic predator feeding on unattended young or infirm individuals. But in addition to the bony remains some curious elongated eggs, laid in pairs, have been found nearby. One of these eggs has been scanned and opened and has revealed part of the embryo of a tiny *Troodon*. Clearly the predator was taking advantage of the relative security of the nesting ground to lay its own clutches of eggs!

New eggs and embryos

Large numbers of eggs have been discovered over the years in Northern China, Mongolia, southern France, North America and South America. As a result, much more is now known about the reproductive behaviour of dinosaurs. While CAT scanners have helped to identify embryo bones inside eggs without damage to the eggs, others have developed more intrusive methods of investigating their contents. A private collector, Mr Terry Manning, with advice from Leicester City Museum has used dilute acid to etch away the surface layers of the egg to reveal the contents of a few of these, and the results have been spectacular. Jack Horner in Montana has had similar success with the eggs of the dinosaur *Orodromeus* and has managed to extract from a large number of eggs, collected from Egg Mountain in recent years, a growth series from tiniest embryo to hatchling.

FOOTPRINTS RECONSIDERED

The last few years have seen renewed interest in dinosaur footprints and trackways. Their history, as we have seen, dates back to the early days of dinosaur research, but their true value has for the most part not been realized. Some have been interesting in their own right, like those discovered by Roland T. Bird of a predator apparently tracking its prey, or a floating sauropod kicking itself along with its front feet. However, in the absence of proof concerning precisely which species of dinosaur left which tracks – dinosaurs never seem to have died and been preserved standing in their tracks – their relevance and interest seemed to be only moderate. Yet tracks are, apart from eggs, the only evidence that we have of what living dinosaurs did during their lifetime.

At the forefront of the revival of interest in dinosaur tracks has been Dr Martin Lockley of the University of Colorado, Denver, who

These parallel row of footprints were left by a group of sauropods as they walked across a moist lowland plain - evidence that these dinosaurs moved around in social groups.

in 1986 described how trackways can be used to greatly enhance our understanding of dinosaurs and their biology.

Trackways provide instant evidence of the posture and style of locomotion of the trackmaker – as in the case of Bird's demonstration that sauropods did not sprawl. The angle at which the feet are held is also characteristic: for example, theropods tend to have footprints which face directly forward, while those of ornithopods tend to be slightly turned in. Trackways may also indicate the speed at which an animal is moving. As might be expected, the majority of tracks are made by walking animals, but there are a few exceptions which show small theropods running very quickly. There is also evidence to suggest that larger theropods may have been capable of moving at 20 or 30 mph (30–40 km/h).

Trackways may also provide indications of the social preferences of dinosaurs. Herding can be indicated by large numbers of prints going the same way, and it may be possible in some instances to deduce herd structure, such as that suggested by Bakker for sauropods moving in a formation to protect their vulnerable young. Indeed, herding and mass footprint sites are beginning to be looked at across large geographic areas to see whether there is any consistent evidence for migratory herding among dinosaurs. This evidence if corroborated could then be linked to the evidence from the nesting colonies, which tends to suggest that ornithopods, at least, migrated.

Trackways are likely to be left under particular environmental conditions. The ground must be fairly soft to show footprints, which gives clues about the ground surface and climate. This information, added to any other known facts, may reveal something useful about the environment. For example, dinosaurs could churn up the ground to an unbelievable extent – a phenomenon known as "dinoturbation." In addition, if footprints of a wide range of animals are discovered in the same place, it may prove possible to build up a picture of the local fauna.

The churning of soil can also have another, less obvious effect, and that is on the plants growing in areas disturbed by herds of trampling dinosaurs. Flowering plants, which first evolved during the Early Cretaceous Period, are often able to reproduce rapidly. They can quickly colonize areas that have been disturbed – like the weeds that spring up in a freshly dug garden. The rise in the number of herbivorous dinosaurs during the latter half of the Jurassic and into the early Cretaceous may well have made severe inroads on the non-flowering plants which formed the bulk of their diet. Dinosaurs' trampling of the soil may well have provided ideal conditions for the spread of quicker-reproducing flowering plants, and hence the evolution of new types. In that case dinosaurs would be responsible for some of the diversity of plants that we see around us today.

The evolution of dinosaurs can also be traced through footprints. For example, Middle Triassic dinosaur-like footprints have been discovered. Such tracks are very rare – only about 3 per cent of known tracks at this time – and indicate perhaps the very earliest and rarest types, of which we as yet have no skeletal fossils. It has also proved possible to chart the increase in the number of sauropodomorphs in the Jurassic, and the later abundance of ornithopods in the Late Cretaceous Period, quite independently of the body fossil evidence. Such correspondences increase researchers' confidence in their attempts to look at larger-scale evolutionary change.

DID THEY MIGRATE?

During 1985 dinosaur bones were discovered in Late Cretaceous sediments on the North Slope of Alaska by a team of researchers from the University of California at Berkeley, headed by Professor Bill Clemens. The bones were mainly those of hadrosaurs, but a few teeth of tyrannosaurs and troodons were also found. Then in 1989 a team of researchers working for the British Antarctic survey, and including Dr Jerry Hooker from the Natural History Museum in London, discovered the partial skeleton of an ornithopod dinosaur in Antarctica. Again the rocks were of Late Cretaceous age. This brings the distribution of dinosaurs incredibly close to both the Arctic and Antarctic Circles, and again raises the question of the physiology of dinosaurs – did they need to be "warm-blooded" to survive such northerly or southerly conditions?

Even though there were no ice-capped poles at this time, conditions on land this far north and south simply must have been more seasonal. The winter periods in both hemispheres would have been long, with prolonged periods of darkness, while the summers would have had very long day lengths. The conditions in winter would have been crucial for the survival of any animals at these latitudes because the prolonged darkness could have caused lower temperatures and serious reductions in the food supply, particularly for the herbivore.

Could dinosaurs overwinter in such conditions? Must they therefore have been endothermic like mammals in order to survive? Two crucial questions need to be answered. The first is whether or not the animals living at such high latitudes were permanent residents. They might have visited only in summer. The Berkeley team have found bones of juvenile as well as adult hadrosaurs at their site, which might indicate that the creatures were resident. However, to date no nest sites or very young animals have been discovered and, as Dr Robin Reid of The Queen's University, Belfast, has pointed out, the smaller individuals might simply represent the minimum size at which animals could join herds migrating to this area.

The second question relates directly to the endothermy debate and is concerned with insulation. High-latitude dinosaurs would have needed significant insulation to prevent heat loss through the skin in cool or cold conditions. There is no evidence for any external fur on the skin of hadrosaurs, which would have been vital to reduce wind chill. An insulating layer of blubber is found only in marine mammals. The evidence, meager though it is, seems to suggest that these dinosaurs were migrants.

Additional evidence in favour of the latter view comes from the plants which grew in the Alaskan area at this time. The British researcher Dr Robert Spicer of Oxford University has collected a very rich flora from this area which indicates that the plants were quite lush and subtropical in many respects. However, he has also been able to show that some forms – even including cycads – were deciduous, losing their foliage in winter. The implication is that the feeding grounds for dinosaurs such as hadrosaurs would have been barren or nearly so dur-

ing the winter, but rich and varied in summer. This points very strongly towards the dinosaurs found here being migrants. It seems likely that the predators discovered among the hadrosaur remains were tracking the large herds of ornithopods in the way that wolves follow herds of reindeer.

Footprint evidence is beginning to be looked at for evidence of mass migrations between north and south along the margins of the mid-continental seaway which divided North America down the middle during the Late Cretaceous.

Hadrosaurs may have migrated in large herds north and south following seasons of the year. This small herd of Telmatosaurus *lived in Central Europe during the Late Cretaceous.*

ABOVE: *Fine fossil preservation means that extremely accurate and life-like restorations can be made of* Archaeopteryx, *such as this model made by Arril Johnson for the City of Bristol Museum.*
RIGHT: *The exceptional quality of preservation in lithographic limestone can be seen in the very latest discovery of an* Archaeopteryx *specimen from Solnhofen.*

CHAPTER SIX

The Origin of Birds

Whether dinosaurs and birds are related has, as we have already seen, been much debated. Thomas Huxley first proposed that dinosaurs and birds shared a common ancestry, in lectures which he published in 1868 and 1870. These views were based on his recognition of the bird-like legs of some dinosaurs, and the small size and very bird-like appearance of the Jurassic dinosaur *Compsognathus*. However, these views were rapidly overtaken by events. Many new, and decidedly non-bird-like, dinosaurs were discovered over the next two decades. Professor Harry Seeley (see page 50) was able to show that dinosaurs belonged not to one but apparently to two distinct groups, the Ornithischia and Saurischia. This made it harder to make the flat claim that birds were derived from dinosaurs, for there was now the question of which group. The final handful of nails in the coffin of Huxley's idea was provided by Gerhard Heilmann in his detailed book *The Origin of Birds*, published in 1926, in which he seemed to have proved that dinosaurs, however bird-like, could not be considered the ancestors of birds since they lacked the collar bones from which birds derive their wishbone.

Until 1973 Heilmann's views held sway among paleontologists and zoologists alike. If some dinosaurs seemed to resemble birds, it was an example of evolutionary convergence rather than an indication of true kinship.

By a quirk of fate a small area of southern Germany has provided a central role in revealing the origins of birds. This area, near the Bavarian town of Solnhofen, contains several limestone quarries. The limestone is of exceptionally fine quality and, unusually for this type of stone, can be split into thin, smooth sheets. These can be used to make lithographic stones for printing illustrations.

The fine-grained limestone was originally laid down in a very calm tidal lagoon where the ebb and flow of water caused thin, even layers of lime to build up on the lagoon bottom. Any animals and plants which died and were washed into the lagoon would be compressed within these fine limestone layers and preserved as fossils. As a result the fossils which are sometimes exposed as the slabs are split by quarrymen are preserved in exquisite detail. Birds, being mostly light and fragile creatures, do not preserve well as fossils and are very rarely found. Even rarer, however, is the preservation of impressions of their feathers; yet even these have been preserved on lithographic limestone.

The key to the origins of birds is at present one particular species, *Archaeopteryx lithographica* (the first name means "ancient wing").

THE FIRST BIRD

The first evidence of *Archaeopteryx* to be discovered was a single beautifully preserved feather found in a quarry near Solnhofen in 1861. It caused something of a sensation of the time. Until this discovery was made birds had been thought to have a very recent fossil history, because their remains (other than the Revd Professor Edward Hitchcock's Triassic bird footprints from the Connecticut Valley – see page 67) are very rarely found and had only been recovered from much newer rock of the Cenozoic Era. This single discovery was made in rocks of Upper Jurassic age, and so moved the history of birds back well over 100 million years! The importance of this single feather can be appreciated by the fact that it was scientifically described and in its own right given the scientific name of *Archaeopteryx lithographica* by Hermann von Meyer the following year. Soon afterwards quarry workers discovered a nearly complete skeleton of a small feathered animal in the same quarries.

The timing of the discovery of these specimens could not have been better from the point of view of evolutionary biologists. It was barely two years since Charles Darwin's book *On the Origin of Species* had been published. In this great work he put forward an explanation of how species may have changed or evolved in time; his theory involved the mechanism of natural selection, which required great periods of time to operate; it also predicted the likely appearance in the fossil record of intermediate animals, which might bridge the anatomical gap between what are today quite distinct types of animals. *Archaeopteryx* possessed features which were both bird-like and reptile-like, and seemed to be a genuine "missing link".

The excitement which this second discovery created brought the specimen to the attention of the district medical officer for the quarry, Dr Karl Hüberlein. He obtained the fossil in exchange for the medical fees he was owed, and offered a collection of specimens – including *Archaeopteryx* – for sale to any museum at the unprecedented price of £700 ($3,500 at 1861 values). Prospective customers were invited to examine the specimen of the extraordinary creature, but neither notes nor sketches were to be made.

Andreas Wagner, a noted anti-Darwinian from Munich, provided one of the earliest descriptions of the creature on the basis of a certain amount of hearsay and borrowed sketches done from memory. He noted that the animal combined feathers with a reptile's long bony tail and clawed fingers. He therefore suggested that this animal was a reptile which had independently acquired feathers, and called it *Griphosaurus*, a name taken from the legend of the griffin or gryphon, a monster with the head and wings of an eagle and the body of a lion. Wagner was most anxious to forestall any evolutionary interpretation of this fossil, so he was at pains to promote the idea that its feathers were nothing whatever to do with the feathers of true birds.

Owen's description

The publicity regarding the intriguing features of this animal not surprisingly attracted the attention of Richard Owen, who was at this time superintendent of the natural history collections of the British Museum. The keeper of the geological collections, George Waterhouse, was sent to attempt to buy the Solnhofen collection for the museum, which he eventually did for the original asking price. Within a few months of its arrival in Britain Owen had described *Archaeopteryx* in a scientific paper presented to the Royal Society.

The specimen shows a beautiful long tail, down the centre of which runs a row of bones – typical of a reptile, for a bird has a short stumpy tail forming the "pope's nose." Fringing the tail on either side is a fan of clearly preserved tail feathers. The legs are long and slender, ending in bird-like feet. Near the front of the chest is a strong and well preserved furcula (wishbone), again pointing very clearly to the creature being a bird; and the wings display a fine array of primary and secondary feathers as do those of any bird. In contrast with birds, the hand has three well developed, sharply clawed fingers. The one thing missing from this specimen was the head, which might have provided firmer evidence of the bird-like nature of the creature.

Owen concluded that this was definitely an ancient bird, and one that was extremely interesting because it showed a variety of primitive vertebrate features which had become modified in later birds.

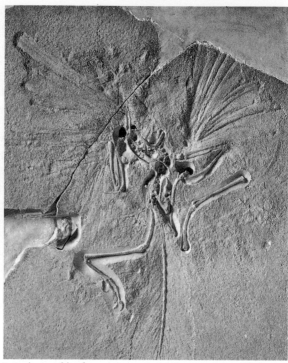

ABOVE: *The very first specimen of* Archaeopteryx *to be discovered was this beautifully preserved feather, found at Solnhofen in 1861.* RIGHT: *Not long afterwards an almost complete feathered skeleton was found. This was bought by the Natural History Museum in London and described by Professor Richard Owen.*

Thomas Huxley's interpretation

Owen's great adversary in the dispute on evolution, Thomas Huxley, did not see the need to overemphasize either the bird or the reptile features of *Archaeopteryx*. He was content to point out that Owen had made some elementary anatomical mistakes when describing this fossil, just as he was able to demonstrate the creature had both reptile and bird features. Huxley was, however, rather lukewarm over its true value as an evolutionary intermediate, because he was convinced that birds had evolved on Earth much earlier than the Jurassic. He had been particularly convinced by Hitchcock's footprint evidence of large Triassic ground-running birds. This made *Archaeopteryx* a rather late member of the bird group and therefore interesting, but only as an intermediate form.

More important, as far as Huxley was concerned, was the evident general similarity between birds and dinosaurs provided by animals such as the small reptile *Compsognathus*, which had also been found at Solnhofen. This could be used as evidence of the bird-like nature of dinosaurs, as opposed to their mammal-like affinities which had been proposed by Owen.

More finds

In 1876 another specimen of *Archaeopteryx* was discovered in the same area of Germany. This specimen was even better preserved than the first skeleton and included a very fine head at the front of a perfectly articulated body, and wonderfully preserved plumage. It now resides in the collections of the Berlin Museum for Natural History. The new specimen confirmed the reptile-bird nature of the animal, for the jaws were not rimmed with a horny bird's beak but were edged by small, sharp teeth like those of a reptile.

Since 1861, a total of seven skeletons, in addition to the original feather, have been recovered from this area of Germany. Some are complete, others fragmentary; and the most recent one has been described by Dr Peter Wellnhofer of the Bavarian State Museum of Natural History, in 1993, as a new species *Archaeopteryx bavarica* ("the ancient wing from Bavaria").

The Berlin specimen of Archaeopteryx, discovered in 1876, is beautifully preserved and has a fine skull. The detail is much finer than the 1861 specimen and has considerably advanced our knowledge of the link between early birds and dinosaurs. It proved conclusively that this early feathered creature had teeth in its jaws.

THE SEARCH CONTINUES

Arguments about the origin of birds from reptiles have been going on for a long time. Though centered on *Archaeopteryx*, they have involved many reptiles.

Dinosaurs: 1868 – 1926

For two decades after Huxley's 1868 lecture there were several strong advocates of dinosaur origins for birds. However, as the range and variety of dinosaurs increased, so the apparent similarity between them and birds decreased. The theory was, it seemed, fatally damaged by Heilmann's work in 1926. Ironically, the rejection of an ancestry which had once been seen as central to the Darwinian mechanism of evolution was made possible by two evolutionary propositions which themselves had sprung from Darwinism. First, Dollo's Law, which states that evolution is irreversible, made it impossible for birds to have evolved from ancestors without collar bones; and second, the anatomical resemblances between dinosaurs and birds which Huxley (one of Darwin's most important supporters) and others found so convincing could be explained as a product of evolutionary convergence toward a similar way of life (see page 81).

Archosaurs: 1926 – 1970

Heilmann selected a more general group of reptiles, which were ancestors of dinosaurs, crocodiles, pterosaurs, and several other groups, and which are usually called archosaurs. They were abundant during the Triassic Period. Some, such as *Euparkeria*, seem relatively light and agile and, since they have collar bones, are not disqualified for that reason from bird ancestry. The choice of such early creatures as ancestors also suggested that birds go back farther than *Archaeopteryx*, as Huxley had believed. This raised the possibility that *Archaeopteryx* may have been an early evolutionary offshoot from the main line of birds, and perhaps not as typically primitive as was first thought.

Heilmann thought Triassic creatures such as Ornithosuchus, *a close relative of* Euparkeria, *were the likely ancestors of birds.*

Ornithischian dinosaurs: 1970

After a prolonged period during which Heilmann's views remained unchallenged, Dr Peter Galton of the University of Bridgeport, Connecticut, proposed a dinosaur ancestry for birds once again. However, rather than proposing a general dinosaurian ancestry, Galton chose a specific group, and an obvious one at that; the ornithischians – the dinosaurs which have a bird-like pelvis.

Galton was in fact reviving a theory which had been roundly dismissed by Heilmann back in the 1920s. If his argument is read carefully, it is clear that Galton was being extremely guarded in his proposals. He recognized that no known ornithischians are suitable as bird ancestors, not least because they are all herbivores and show specializations which are inconsistent with bird anatomy. So to justify his claim he pushed bird ancestry back to a time when the very first dinosaurs were evolving, and was really proposing an advanced archosaur or "protodinosaur" with a bird-like pelvis as the original bird ancestor. When Galton's theory is reduced to this level of argument, it is clear that it is not a great deal different from Heilmann's; and within a few years Galton was to admit that his theory was probably not correct. The fossil record shows no clear evidence for a bird-like protodinosaur.

Crocodiles: 1972

A British paleontologist, Dr Alick Walker of the University of Newcastle, in 1972 proposed that modern birds were more closely related to a group of Triassic crocodiles. He had been involved in a detailed study of the Triassic crocodile *Sphenosuchus*, and was able to point to a number of unexpected similarities in the form and arrangement of the skull bones in birds and this fossil. This provoked him to look in greater detail at the structure of living birds and crocodiles. Numerous similarities were indeed brought to light, in the structure, fore limbs, and ankles of embryonic birds and crocodiles. His principal suggestion based on this careful work was that the ancestors of birds and crocodiles seem to have adopted one of two ways of life. One group of rather slender, lightly built crocodile-like creatures adopted the habit of tree climbing, and ultimately became birds; while the other became larger amphibious types and developed into what we would now regard as typical crocodiles.

Fascinating though much of this work was, it was curious that Walker chose not to use *Archaeopteryx* in the comparisons he was making between crocodiles and birds. Despite this, his theory has attracted some support among paleontologists. Again, it can be seen as a subtle modification of the suggestions of Gerhard Heilmann.

Dinosaurs revisited: 1973 onwards

During the 1960s Professor John Ostrom was deeply involved in the detailed anatomy and description of the remarkable new theropod

Deinonychus (see page 123). The unusual nature of this dinosaur made it necessary for him to study the few small theropods held in the collections of museums around the world. Among these dinosaurs was the fine specimen of *Compsognathus*, one of the very few small theropod dinosaurs to have been found with its limbs intact, which had been collected from the same quarries that had yielded *Archaeopteryx.*

Ostrom was apparently attracted to *Archaeopteryx* by his discovery of a previously unrecognized specimen. It was a part of a leg embedded in a slab of lithographic limestone, and had been thought to be from a flying reptile (pterosaur). Ostrom spotted the impressions of some feathers on the slab and realized that this must be *Archaeopteryx.* However, he also noticed how much the leg bones resembled those of a theropod. This chance discovery alerted him not only to *Archaeopteryx* itself, but also to the similarity of its skeleton to that of a dinosaur.

During the late 1960s and early 1970s Ostrom examined all the then known specimens of *Archaeopteryx.* Meanwhile yet another, this time nearly complete, specimen turned up in a German museum collection at Eichstatt. Discovered in 1951, it has the distinction of having been classified as the dinosaur *Compsognathus* for nearly twenty years! The slab on which it lay showed practically no evidence of feather impressions.

FROM DINOSAURS TO BIRDS

The small theropod dinosaur Compsognathus *illustrated below shows many of the bird-like qualities of small carnivorous dinosaurs. The head is low and pointed, the eye sockets are large, the neck is long, slender and curved and the rest of the body is delicately built. Note that the arms are rather short compared to birds.* Archaeopteryx *(below left) displays an amazing combination of dinosaur and bird characters. The jaws lined with teeth resemble* Compsognathus, *as does the remainder of the skeleton, except for the long arms, pelvis and feathers,* Corvus *(a modern crow - below right) shows the toothless jaws, large breast bone and short tail typical of modern birds.*

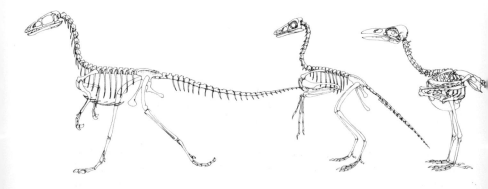

Ostrom spent much of his time repeating the work of Heilmann, but if anything in greater detail and supplemented by considerably greater paleontological information. He was able to list a great number of detailed similarities – twenty-one in all – between small theropods and *Archaeopteryx*, in the fine structure of the bones of the skull, arms, shoulders, back, legs, hips, and tail. The arms and legs of theropods are remarkably similar to the limbs of *Archaeopteryx*. The one overwhelming obstacle which Heilmann was unable to overcome in his time, the absence of collar bones, had been removed by the discovery of small, strap-like collar bones in a number of theropods in the late 1960s and early 1970s, mainly as a result of new discoveries made by the Polish-Mongolian expeditions in Mongolia. This allowed Ostrom to be far freer to pursue the arguments developed by Heilmann. In one review he wrote:

> It is more probable that *Archaeopteryx* acquired the large number of derived "theropod" characters by convergence or in parallel at the same time that these same features were being acquired by some coelurosaurian theropods – presumably from a common ancestor? Or is it more likely that these many derived characters are common to some small theropods and *Archaeopteryx* because *Archaeopteryx* evolved directly from such a theropod? There is no question in my mind that the last explanation is far more probable.

Take your pick

Any final judgment on the matter is based on personal preference for a particular idea or theory. Several groups of workers have pointed out problems with Ostrom's views concerning the closeness of the relationship between dinosaurs and birds. Fine points in the anatomy of the ear in birds and crocodiles have been put forward as persuasive evidence that the ancestry of birds has to be put back into the Triassic – supporting the views of Alick Walker. Yet others favour an archosaur ancestry of birds because of fine differences seen in the structure of the shoulder, wrist and pelvis.

In addition to the work of Professor Ostrom, detailed studies of the anatomy of theropods as a whole by Dr Jacques Gauthier of the California Academy of Sciences have pointed to a very strong similarity between *Deinonychus* and its relatives (the deinonychosaurs, or dromaeosaurs as the group is sometimes called) and *Archaeopteryx*.

At the present time the majority of dinosaur paleontologists (myself included) favor a shared ancestry for theropod dinosaurs and birds probably in the Early Jurassic Period, and therefore support Ostrom's conclusions.

A late entrant

As an indication of the fact that ideas may yet change there is the case of "*Protoavis*" ("first bird" – a provisional name). Its discovery was announced by Dr Sankar Chatterjee of the University of Texas, Lubbock, several years ago, but it has not yet been scientifically described. It is a fragmentary skeleton of a small, lightly built reptile, which does not appear to have been preserved with feather impressions. As a result the suggestions concerning its bird affinities must relate to its bony anatomy alone.

The specimen dates from the Late Triassic. It is perfectly reasonable to believe that this may indeed be an extremely early fossil bird, but such an opinion must be expressed very cautiously. There were a number of very small, lightly built archosaurs living in the latest Triassic, and some may indeed have been quite bird-like in outward appearance. This was, after all, a time when the first pterosaurs evolved; these flying reptiles, although their origins are obscure, must have arisen from extremely lightly built, tree-dwelling archosaurs. There may well have been a considerable diversity of small, insect-eating, tree-dwelling archosaurs at this time, during which the ancestors of birds evolved in parallel with the pterosaurs. Since the anatomy of their wings is totally different, there can be no question of a close relationship.

The critical questions that will need answering as far as "*Protoavis*" is concerned are: first, whether this creature is a highly specialized archosaur or an early dinosaur; and second, having established that answer, whether the material is sufficiently well preserved to tell if it is really a bird ancestor or a genuine early bird. Whether the specimen is well enough preserved to be able to answer all or any of these questions remains to be seen.

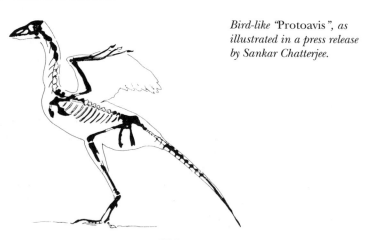

*Bird-like "*Protoavis*", as illustrated in a press release by Sankar Chatterjee.*

LEARNING TO FLY

How any creature can evolve the power of flight has long been a puzzle. The arguments about birds date back to two theories put forward by Samuel Wendell Williston and Othniel Charles Marsh in 1879 and 1880 respectively. Their theories, and later ones, can be summed up by the two phrases "ground upwards" and "tree downwards."

Ground upwards

The belief which underlies this theory is that the ancestors of birds were, like small theropods, fast-running animals. The idea, developed by Williston and repeated by Franz Baron Nopcsa in the 1920s, was that wings – or rather wing-like arms – first made their appearance in the form of propellers. The animal would flap its arms when running to increase its speed along the ground. This strategy would have been successful, so that the animal evolved larger and larger propellers. Eventually these would be so large that they could lift the creature off the ground for a short hop. From there they evolved into real wings. This scenario seems to be supported by observations that can be made among birds living today. Some of the larger and heavier birds, such as geese, swans, and albatrosses, frequently use a long run-up, vigorously flapping their wings, in their efforts to get airborne. The earliest birds would also have had to run fast to gain sufficient ground speed to take off with their inefficient wings.

While this might at first sight seem to be a reasonable idea, the analogy with an aircraft propeller and its transformation into a wing is misleading. Most obviously arms, which cannot spin like a true propeller, cannot produce a steady thrust. Early "proto-wings" would have to be flapped up and down. This presents a physical problem for our theoretical sprinter; once arms or proto-wings are stuck out from the sides of the body, the body's smooth streamlining is broken, so the wings would create a combination of aerodynamic lift and drag both of which would actually slow the creature down! Drag would act like an air brake to slow the animal down directly. Lift, although ideal for flight, would be a great inconvenience to a ground runner since it would reduce the weight of the animal and spoil the grip which its feet had on the ground.

It thus seems that the comparison which was drawn between early birds and modern birds which run before takeoff is misleading. Modern birds run to reach a suitable takeoff speed because they already have fully functional wings with perfect aerodynamic properties for flight. Early birds can be presumed to have wings which had neither perfect functional mechanics (bones, joints and muscles) nor perfect aerodynamics, and the biological advantages of imperfect wings do not seem to be at all apparent.

THEORIES OF FLIGHT

"Ground upwards"

BELOW: *The image of a scampering protobird attempting to become airborne by running at high speed has had many doubters. One novel suggestion has been that flight evolved as an accidental by-product of swinging the feathered arms in order to catch insects.*

"Trees downwards"

ABOVE: *The ability to fly may have be developed by stages which started with simple parachuting to the ground usi feathers to help break the fall, followed by the development of feathered wings for gliding, and finally full flight.*

Most paleontologists tend to dismiss this theory because of the simple physical problems of trying to develop flight in a fast runner. This line of reasoning, which tends to suppose that bird ancestors developed feathers and wings because in some way they "knew" that they could be ultimately used for flight, is also extremely suspect.

A novel twist to the "ground upwards" theory was provided by John Ostrom in 1974. As we have seen, Ostrom was struck by the similarity between *Archaeopteryx* and fast-running theropods, and this led him to believe that flight must have originated among small, fast-running ground-dwelling predators. If such creatures were taken to be about the size of *Archaeopteryx*, which in life must have been about the size of a crow but with longer legs, they would very probably have been insectivorous. *Archaeopteryx* has small, spiky teeth, which seem consistent with a diet of insects. Starting from this premiss Ostrom developed a completely different theory on the origin of both feathers and flight.

Ostrom suggested that such highly active insect predators would have pursued their prey at speed, catching flying insects such as large dragonflies (whose fossil remains have been found in the Solnhofen deposits) with leaps and darting movements of the head. Small, highly active creatures such as this may well have been endothermic, in which case they would have needed some degree of insulation.

Small size produces a relatively high ratio of surface area to volume, so that the creature would lose heat very quickly if not properly insulated. It was possible, said Ostrom, that these creatures evolved a feathered edge to their scales, which would have made them retain heat better. However, such developments may well have been of secondary advantage to these predators. Feathery scales, particularly ones fringing the arms, could have been used as a primitive form of insect net to assist with the capture of evasive flying insects. Ostrom therefore imagined early "proto-birds" as ground-dwelling insect predators with feathered fringes to their arms. The next step, according to this theory, is the development of an ability to leap into the air in pursuit of their prey before they flew out of reach. Leaping and thrashing with the feather-fringed arms may well have produced a fluttering sort of flight which would have enabled the "proto-bird" to catch its prey.

Flight in Ostrom's theory is therefore seen not as the ultimate intent of bird ancestors, but as an accidental development – a byproduct of the habit of feeding on insects. The theory also gives a plausible account of the origin of feathers, or at least of scales with feathery fringes, as a way of insulating the body in small endotherms; their development as aids to flying may have occurred considerably later. This alternative "ground upwards" theory has received much more serious consideration than the previous one.

Trees downwards

In 1880 Marsh described the conditions under which he believed that flight may have emerged. Drawing a parallel with animals living today, Marsh proposed that the ancestors of birds may have been tree-dwelling creatures. For creatures living in such a precarious habitat, there might have been very strong evolutionary pressure in favour of any attribute, no matter how seemingly small, which would tend to break the fall of animals if they lost their footing. That might be especially likely if such creatures were attempting to catch and eat agile and fast-moving insects under the forest canopy.

Given this set of conditions it was possible to imagine how ancestral birds might have evolved through the stage of animals which had developed parachute-like structures to break their fall. There are many animals living today including lizards (*Draco*), snakes, various "flying" frogs (in reality gliding frogs), rodents, phalangers, and primates living in the treetops which have developed a wide range of parachute structures: feet that can be spread out like umbrellas (frogs), extensible ribs (*Draco*) and webs of skin which can be stretched out between the front and back legs (rodents, phalangers, and primates).

This theory also suggests how feathery scales could have evolved gradually. They could well have helped to break the fall of small, tree-climbing creatures. Evolution may well have favored those animals which were best able to break their fall, or perhaps to parachute to safety when pursued. Having developed some sort of limited parachuting ability through the development of feathery scales, ancestral birds are then supposed to have gradually perfected the ability to glide from tree to tree.

The saving in energy made possible through gliding from branch to branch, rather than having to climb up and down tree trunks, is obvious. From a gliding stage, the development, or refinement, of gliding into active flapping flight would appear to be another rela-

Gliding between the trees, Archaeopteryx *would have been capable of catching larger, slow-moving insects.*

tively straightforward stage in the exploitation of the air, perhaps in order to catch insects on the wing.

Gliding is relatively easy, and does not require much effort. Active flapping flight is something else – and this is reflected in the fact that there are so few true fliers among the vertebrates (animals with backbones). In the past only pterosaurs seem to have achieved flight, and today only birds and bats are active flapping fliers. The difficulties created by active flying include the requirement for powerful muscles to power the wingbeat, a strong but very light skeleton, a powerful heart and lung system to supply oxygen and food to the flight muscles; and a highly sophisticated sensory system and brain to control and adjust the flight path of the animal at all times. This is a daunting list of requirements, and all must be provided simultaneously if the "flying machine" is to work.

Heilmann's "Proavis" clambers among the branches ready to launch itself on a short, free-fall leap to the next tree.

Despite these difficulties, Marsh's theory became widely accepted, and was in a sense brought to reality by Gerhard Heilmann in his book *The Origin of Birds*. Heilmann reassessed Marsh's proposals and concluded that this did seem to be the most logical mode of evolution of flight in birds. To reflect this, Heilmann (also an artist) drew a very fine illustration of a purely hypothetical bird ancestor: his *"Proavis"* (not to be confused with *"Protoavis"*), a lightly feathered creature which he pictured scrambling in a tree.

Up or down?

The origin of feathers seems relatively uncontroversial: both "up" and "down" theories can accommodate it. It seems perfectly possible, if unprovable, that feathers developed initially as scales with fringes on.

The origins of flight are not quite as clear, but I must admit to favoring the "trees-down" theory. In general terms the origin of flight seems more probable from within a group of small, tree-dwelling creatures. The reasons for preferring this are as follows. Takeoff is far easier for an animal which launches itself from the branch of a tree. Using gravity to create speed, and the body contours to create some

206

lift during descent so that gliding becomes possible, seems inherently more logical, and is supported by the wide range of animals living today which do so. The transition to active powerful flight is admittedly a difficult stage. However, these hypothetical tree dwellers would have had one advantage: the part which had become specialized as a "parachute" was their muscular and mobile arms. This, of course, is also the case with pterosaurs and bats. The other gliding vertebrates used parts of their body which were less suitable for development, and have not managed to do anything more than steer.

Could Archaeopteryx fly?

There has been much debate concerning the flying ability of *Archaeopteryx*. One of the most obvious reasons for the debate is the absence in all specimens of *Archaeopteryx* of a bony breastbone or sternum. This is the large shield-shaped piece of bone in the centre of the breast of a bird, which has a prominent midline ridge to which the large flight muscles are attached. The argument is simple. Modern birds have huge flight muscles attached to the breast bone; *Archaeopteryx* did not have a large bony breast bone; its flight muscles were at best weak and its flight must have been relatively feeble by comparison with modern birds.

The usual picture of *Archaeopteryx* is of a simple tree-climbing creature, with the ability to glide from tree to tree, perhaps assisted by the occasional flap of its wings. Recent work, however, suggests that this view of the ancient bird may be a little unfair. Examination of the feather pattern reveals that it has wings with an arrangement of primary and secondary feathers much like those of a modern bird. The primary feathers are the longer feathers attached to the end of the wing (to the bird equivalent of the hand) and provide much of the forward propulsion during flight. The secondary feathers are attached along the remainder of the arm and tend to be a little shorter than the primary ones; they are responsible for lift.

Furthermore, the Canadian paleontologist Dr Alan Feduccia has shown that individual feathers of *Archaeopteryx* are asymmetrical. That is to say that, when viewed from above or below, the vane of the feather is not divided into two equal halves by the central stalk (rhachis). The front half of the vane is relatively narrow, and the rear half is much broader. This pattern is exactly the same as is seen in all flying birds today, and makes the feather aerodynamically efficient.

Both pieces of evidence point to *Archaeopteryx* being an active flier. The more serious observation, the lack of a large bony breastbone, has been countered by the careful study of the flight muscles of modern birds. The furcula (wishbone) was prominent in *Archaeopteryx*, and far more stout than in living birds of comparable size. The prime

Although not very close relatives of birds, pterosaurs such as Dimorphodon *were able flapping fliers.*

wing elevators are attached to this bone in modern birds, and this is good circumstantial evidence that at least part of the typical bird system of flight muscles was well developed. It has also been pointed out that bats do not have such well developed breastbones as birds. Their large, powerful breast muscles are tied together at the midline with tough ligamentous sheets.

It is now thought that *Archaeopteryx* was a reasonably competent flier, capable of flapping its wings quite vigorously. It was certainly unlikely to have been a mere glider, as has been envisaged in the past. The long bony tail would have been a useful rudder, and would have made *Archaeopteryx* a very stable flier – though that means that it would not have been able to perform great feats of aerobatics. And indeed its skeleton was generally heavier than a modern bird's, suggesting that it did not have great aerial skills.

OTHER EARLY BIRDS

Whatever its flying ability, and whatever its ancestry, *Archaeopteryx* is a marvelous fossil, providing a window on a moment in the evolutionary history of birds. Paleontologists need many more such windows if they are ever to decipher the evolutionary history of birds, but unfortunately we cannot expect much help from the fossil record in this respect. Not only are birds among the lightest and most fragile of vertebrates, but they live in environments which are only very rarely likely to lead to their preservation as fossils.

A new bird skeleton has recently been discovered in China, which appears to date from the Early Cretaceous, and is therefore the second oldest of all known birds. It is hoped that this new specimen will help to shed a little more light on the evolutionary history of birds. Beyond that, several birds are known from the Late Cretaceous, both flying and swimming types, which seems to suggest that they became diverse very early in their history. The Late Cretaceous birds retained teeth in their jaws, but teeth were soon lost in the Early Cenozoic.

As birds seem to have been quite close relatives of at least one group of dinosaurs, their survival through to the present day is of considerable interest. In their biology and behavior we may be able to detect traces of their dinosaurian forebears. But birds are clearly different from dinosaurs – even from theropods. The testimony to that fact is that birds survived the extinctions which brought an end to the dinosaurs 66 million years ago at the close of the Cretaceous Period.

ABOVE: *Competing theories suggest causes that range from climatic change and evolutionary
exhaustion to mysterious forces operating in the universe at large.*
RIGHT: *Individual dinosaurs, such as* Dromiceomimus, *and whole species of dinosaurs died
out throughout the Mesozoic Era, but at the end of the Cretaceous Period the whole range of
dinosaurs died simultaneously – why?*

CHAPTER SEVEN

The End of the Dinosaurs

One part of the history of the dinosaurs which has long fascinated people is their extinction. It seems that one of the best known facts about dinosaurs is that they died out 66 million years ago. That moment in time marked the end of the Cretaceous Period and of the Mesozoic Era, and the onset of the Cenozoic Era, which was to be dominated by the rise of mammals and, much later, of ourselves.

The fact of the demise of the dinosaurs has very often been tied with ideas that they were in some way inferior creatures. That is an oversimple view; not only the dinosaurs but many other groups of animals became extinct at about the same time.

Looking simply at the question of the dinosaurs, it must be remembered that dinosaurs of one kind or another lasted on Earth for at least 155 million years; this is a staggeringly long period of time by any reckoning – ranging from the first, rare forms found in the Late Triassic to those that lived at the very end of the Cretaceous. Throughout that immensity of time dinosaur species of great variety evolved and died out, as can be seen from the family tree and time charts on pages 62–63.

It was Georges Cuvier (see page 67) who first demonstrated the fact of extinction in the early years of the nineteenth century through his work on fossil elephants. Cuvier did not see extinction in an evolutionary context, but as the periodic elimination of life forms which were then replaced by the creative power of God.

Rather than thinking of extinction in purely negative terms of death or destruction, it can actually be seen to be a positive force working for the good of life on Earth. Extinction is part of the development of evolutionary diversity. It is essential to evolutionary change, creating a "space" into which new organisms can develop and diversify. A world in which there was no extinction would be a static, unchanging place. Everything that had appeared at the beginning of time would continue endlessly. There would be no place and no space for new varieties to appear, and no chance for change or progress to be made. Extinction is therefore vital to progress and development, and the ever-changing face of life on Earth.

Extinction also allows life to respond to changing conditions. We have seen that, as a consequence of plate tectonics, the continents of the Earth have not stayed in one place or configuration. Such changes, even if continental movements are imperceptibly slow by the standards of a human lifetime, result in slow but inexorable shifts in climate as continents drift across latitudinal belts and as the patterns of ocean currents change. Environmental change brought about by these geological factors can lead to the extinction of groups well adapted to previous climates, and their replacement by species better able to cope with the new conditions. In similar manner the breakup or joining together of continents can affect species either through isolation, or by the introduction of new competitors; in either event extinction rates may rise.

Looked at from this viewpoint, the extinction of dinosaurs does not seem so incredible or perplexing. A more appropriate question might be: How did dinosaurs manage to exist for so long in spite of extinction? No particular species of dinosaur lived for the entire length of the reign of the dinosaurs – many came and went throughout that time. Individual species are unlikely to have existed for much longer than 2 or 3 million years, and perhaps in some instances considerably less than that. But each species formed a vital part of the rich tapestry of life during the Mesozoic, and each must have had its own impact on the ecology of the time, as well as an influence on the species which appeared afterwards.

Yet, while extinctions occurred throughout the reign of the dinosaurs, something extraordinary seems to have happened at the close of the Cretaceous; and this brings the question of dinosaurs and extinctions into sharp focus.

By charting the relative abundance or absence of the fossils of shelly brachiopods in rocks of different ages, it is possible to identify the timing of mass extinctions.

Mass extinctions

The end of the Cretaceous Period seems to mark a sudden, widespread extinction of many different organisms – what we today call a mass extinction event. Again, it was Cuvier who first noted the existence of mass deaths of species in the fossil record. He and Alexandre Brongniart charted the succession of rocks and fossils in the area around Paris and noted that whole faunas appeared to have been periodically wiped out and totally replaced. This was the basis for their "catastrophist" views of the history of life: that life progressed by a series of local catastrophes such as sudden floods, which removed species wholesale and allowed new forms to enter from surrounding areas.

Dinosaurs of one kind or another dominated life on land right up to the end of the Cretaceous Period. At the same time other groups of reptiles dominated the air and water. In the air, pterosaurs seem to have dominated for much of the Mesozoic, although toward the close of the Cretaceous a few types of bird began to appear. In the water

213

there were ichthyosaurs, plesiosaurs, and marine crocodiles in considerable abundance as well as the giant seagoing mosasaur lizards.

In addition to the purely reptilian types living during the Mesozoic there was a whole host of other organisms. Ammonites, noted for their beautiful coiled shells now collected by fossil hunters (and distant relatives of the pearly nautilus of today) lived in great numbers in the seas. Chalky plankton existed in such abundance that their bodies have formed huge thicknesses of chalk, which have given their name to the Cretaceous ("chalky") period. A huge diversity of shellfish such as brachiopods and clams inhabited the coastal shallows.

This tremendous variety of life suddenly vanished at the close of the Cretaceous Period. This time of catastrophic change is known as the K–T boundary, for Cretaceous–Tertiary; the K comes from the Greek for chalk, *kreta*. In general terms any land-living animal more than 3 ft (1 m) long became extinct, as did nearly all large marine reptiles, including the marine crocodiles, but excluding the marine turtles. All the ammonites disappeared, as did almost all chalky plankton species and a great many brachiopods and clams. All the flying reptiles vanished, though the birds did not. The curious nature of this mass extinction is not only that it was widespread in the groups that it affected, but it was also selective. Birds survived apparently little affected, as did freshwater crocodiles, mammals of various kinds came through, and bony fish and sharks were seemingly unaffected. The majority of plant species seem not to have registered the disaster which eclipsed so many animal groups, though there were losses of some early flowering plants and there seems to have been a brief, extraordinarily rich flora of ferns just following the K–T event.

ASSORTED THEORIES

It is clear from what I have said above that the question of why dinosaurs in particular became extinct is ill considered. It reflects the considerable general bias towards dinosaurs, and the fact that they were large and interesting. The misplaced emphasis on dinosaurs has provoked a great number of theories attempting to explain the extinction of the dinosaurs alone. These can be divided into a number of general types.

Bad design

One of the most persistent theories put forward to explain dinosaur extinction is that they had simply lived long enough, and it was time for them to go; this is often referred to as "racial senescence" or "world-weariness." The idea is that races of animals have a time of origin, a period of growth, and then a period of decline and final annihilation – just like a human lifetime. Attempts have been made to suggest that dinosaurs showed signs of racial old age in the bizarre structures that appeared in some species at the close of the Cretaceous Period. There are, for example, the enormously thickened heads of pachycephalosaurs, the huge bony frills of ceratopians, and the elaborate crests of hadrosaurs. As we have already seen, these structures are not at all bizarre, but had a specific role in the animals' lives. Dinosaurs even at the close of the Cretaceous showed no sign of racial "old age;" they seem as vigorous as ever, producing many new and different types – hardly the picture of a group slumping into decline.

In a similar vein various potentially fatal bodily disorders have been proposed as the causative agents of their extinction: slipped disks as a consequence of their extremely large size; hormonal disorders for the same reason; excess body heat or social problems causing malformations of their bones during growth; or progressive diminishing brain size resulting in death through stupidity and inability to cope with changing conditions. None of these examples is persuasive. Extremely few dinosaurs show any damage to their vertebrae which might have been caused by slipped disks; and those examples claimed to show hormonal disorders again relate back to forms such as the pachycephalosaurs with their enormously thickened skulls. Diminishing brain size is instantly refuted by Professor Hopson's work which shows that some of the latest types – especially the Late Cretaceous troodontids – had the largest of dinosaurs' brains.

A novel twist to the argument relating hormonal disorders to dinosaur extinction was provided by Dr Heinrich Erben of the University of Bonn in Germany. Study of the egg shell structure of a Late Cretaceous dinosaur seemed to show a progressive thinning of the shells with time. He postulated that the thinning, which was eventually fatal to many eggs, was a sign of stress-induced hormonal disturbance. Stress is known to cause thinning in bird eggs by upsetting the laying bird's hormones. Did this happen to the dinosaurs? Erben's scenarios envisaged dinosaurs living in lush conditions and becoming superabundant, causing overcrowding and stress among the egg-laying females. Such observations have not been seen in other species of dinosaur – and the colonial nesting sites found in Montana and Alberta (see page 18) directly contradict the theory.

Another range of theories relating to flawed design among dinosaurs relates to diet and digestion. Attempts have been made to prove that the plants upon which dinosaurs fed, such as ferns of various types, contained oils which are considered vital to the efficient passage of food through the gut. The disappearance of such plants has been proposed as a reason for the widespread death of dinosaurs

Dinosaur eggs have been used as evidence to suggest that overcrowding and stress caused dinosaur extinction.

through chronic constipation! Alternatively dinosaurs are claimed to have been unable to detect plant biochemical defences: either alkaloid poisons or substances which inhibit or otherwise disable their digestive systems. Again neither of these claims can be seriously entertained in the light of the remarkable variety of digestive systems employed by dinosaurs. And, perhaps more importantly, the plant fossil record does not show such a dramatic shift in floral types at the end of the Cretaceous as to cause dietary problems.

Other creatures

There is also a wealth of ideas on extinction caused by other groups, either directly or indirectly.

Diseases or parasites are frequently thought to have caused the demise of the dinosaurs. The appeal is obvious, and parallels can be drawn with plagues and epidemics from the present day or recent history. However, in the case of disease the effects are always limited in their effect to closely related species: foot-and-mouth disease is limited to cloven-hoofed animals; rabies affects only certain mammals; and bubonic plague (the "Black Death") affected only humans, having been carried by the fleas on black rats without killing either creature. It is extremely improbable that one disease-causing organism would have affected the entire range of dinosaurs. And in any case Nature tends not to operate by total extermination. The natural genetic variability within individuals of a species almost always results in the survival of a few which are resistant to the most virulent of diseases. Looked at from the parasite's viewpoint, it would be an extremely unwise policy to exterminate completely its host organism, because the extinction of one will result in the inevitable death of the other!

Taking the parasite theme a little further, it has been proposed that the rise of mammals toward the end of the Cretaceous may have

216

Caterpillars can defoliate individual plants, but probably not on a large enough scale to cause mass extinctions.

been linked to the decline of dinosaurs. This scenario involves the contemporary small mammals developing an egg-eating habit, and exploiting it so successfully that they caused a rapid decline in the population of dinosaurs, and their eventual extinction. The improbability of this suggestion must be immediately apparent. Is it really likely that all dinosaur species would have fallen victim to egg predators? Indeed it is not entirely certain that all dinosaur species laid eggs. Taking another point of view, it is well known that there are egg predators today, but in no way do they jeopardize the species whose eggs they eat.

Overkill by carnivores is another theory which runs into the same problems. The supremely powerful predatory features of the Late Cretaceous tyrannosaurs have been used as evidence that they were the ultimate killing machines. So effective were they that they caused the extermination of their prey, and in doing so precipitated the extinction of the entire race. Again, this is an extremely improbable scenario.

Finally in this section I will mention an equally bizarre, but rather more subtle argument. This concerns the evolution of caterpillars, the larval stages of butterflies and moths. The idea is that caterpillars evolved toward the close of the Cretaceous and spread spectacularly. Their immediate effect was to strip the leaves from the plants on which the herbivorous dinosaurs depended. As a consequence the herbivores starved and died, and the carnivores succumbed following the disappearance of their prey. The idea that caterpillars could have caused defoliation of all plants worldwide needs an enormous leap of the imagination. In addition the fossil record of Lepidoptera (butterflies and moths) is so poor that, even though they may have originated in the Early Jurassic, it is not certain that there were significant numbers of these insects around the Late Cretaceous.

Global catastrophes

The implausibility of the above theories which concern dinosaurs alone has given rise to a wide range of explanations based on disruption or change to the environment.

As long ago as 1841, Richard Owen proposed that during the Mesozoic the air had higher levels of carbon dioxide and lower levels of oxygen than those of today – conditions which he thought suited the reptiles best. The end of the reign of dinosaurs was seen as a period of change with oxygen levels rising and carbon dioxide levels falling to those seen today. These changed conditions suited the mammals and birds, but did not permit the survival of dinosaurs. Numerous variations on this general theme have been proposed since Owen's time. Scenarios involving climates becoming hotter or cooler, wetter or drier, have all been proposed at one time or another and linked to the extinction of dinosaurs. Two particularly topical suggestions which have some relevance to our present concerns about global environmental change concern the destruction of the ozone layer and the global warming attributed to the "greenhouse effect."

Destruction of the ozone layer has been linked to great volcanic activity at the end of the Cretaceous Period. The eruptions would have released large quantities of hydrochloric acid into the atmosphere. The acid would be split by sunlight to release chlorine, a highly reactive gas which would have destroyed or severely depleted the ozone layer around the Earth. Removal of the ozone screen would have allowed ultraviolet radiation from the sun to destroy the animals living on land and the plankton in the upper layers of the sea.

The greenhouse theory can be combined with the previous one. Volcanic activity also produces vast quantities of carbon dioxide, and if this occurred at an unprecedented scale at the end of the Cretaceous it could have caused a rapid rise in global temperature. Carbon dioxide is described as a "greenhouse gas" because it allows solar energy to enter the Earth's atmosphere, but restricts the reflection of that energy back out into space. The Earth rapidly heats up under these conditions, creating temperatures which become intolerable for many species. This effect would be compounded by the death of temperature-sensitive plankton, which have a vital role in the conversion of carbon dioxide into oxygen by the ordinary plant process of photosynthesis.

Another theory concerns the Earth' magnetic poles. Geophysical surveys which have charted the patterns of magnetism in rocks on the sea floor have revealed that the Earth's magnetic poles have flipped, so that magnetic north becomes south, and south becomes north, on a number of occasions in the geological past. For example, about three-quarters of a million years ago the magnetic north pole, which was over Antarctica, appears to have changed quite suddenly to its position today. A number of magnetic reversals have been found in rocks laid down in the time leading to the extinction of the dinosaurs. Some physicists believe that the reversal results in a momentary collapse of the Earth's magnetic field as the poles flip over. This would, at least theoretically, allow cosmic rays and other radiation from space, normally deflected by the magnetic field, to penetrate the Earth. Whether such particles or rays would have been lethal in themselves, or could have caused disastrous mutations in certain animals, is unclear.

Supernovae have proved popular among some scientists in the past as the cause of dinosaur extinction. Such explosions mark the violent death throes of a large star which has run out of fuel and

Comets are quite regular visitors to our part of the Universe. If a comet of a similar size to Halley's Comet had dislodged from the Oort cloud and collided with the Earth the environmental consequences would have been disastrous. The possibilities range from a nuclear fireball enveloping the Earth to global cyanide poisoning. At the very least such an impact would have caused rapid global warming, wiping out all forms of life which could not adapt to this sudden change.

collapsed. The Crab Nebula in our own Galaxy represents the debris left after a supernova which was seen on Earth in the year 1054. The supernova which caused the Crab Nebula was not close enough to Earth to have had any perceptible effect. However, if such a supernova had occurred nearby, its effects, such as a flood of intense radiation and an enormous magnetic shock wave, could have severely disrupted life on Earth. But there is no convincing astronomical evidence of such an explosion near enough to our own solar system.

Comets have also recently joined the list of candidates in the search for the cause of dinosaur extinctions. A Nobel prizewinning geochemist, Professor Harold Urey, suggested that a comet the size of Halley's comet (a couple of miles across), if it had collided with the Earth, would have raised global temperatures very rapidly. Indeed, the impact might have set off a large nuclear explosion.

An alternative twist to the idea of a cometary impact comes from the observation that some comets seem to have a crystalline "head" in which significant quantities of cyanide have been detected. A large comet would therefore have caused mass poisoning of the planet.

Scientists' pet theories

Not surprisingly, scientists have tended to put forward extinction theories based on their own expertise. Indeed, it seems that scientists from almost any discipline will throw their hats into the ring. In general this is a very healthy state of affairs: it is clearly good to have dialog between scientists of all varieties. However, such a willingness to propose new theories does hold considerable risks for those concerned, especially if they lack much paleontological knowledge.

One theory put forward by a visual physiologist a few years ago was that dinosaurs became extinct because they were blinded by cataracts caused by excessive ultraviolet light. This theory originated from some work done on the characteristics of the corneas of modern reptiles.

Another theory, advanced quite recently by a plant biochemist, suggested that dinosaurs became extinct because they were unable to taste the alkaloid poisons produced by plants in the Late Mesozoic. Again the origin of this theory can be found in some work done with living reptiles, in particular on the inability of tortoises to detect alkaloids in plants which they were fed under experimental conditions.

In neither case is the theory which has been put forward one which deserves much consideration. Modern reptiles, particularly tortoises, are poor subjects for biological comparison with dinosaurs. It is quite reasonable to say that we human beings are marginally more closely related to dinosaurs than are the tortoises!

However, the point I am trying to make is not the simple destruction of two more theories. It is that both proposals were made by well-respected scientists, who have no doubt done much good work in their own respective fields of research. The universal fascination with the question of the extinction of dinosaurs has led these men to throw their own pet theories into the arena. This usually simply adds to the confusion rather than helping to clarify issues, and seldom adds much to their scientific reputation.

I would now like to concentrate on the main issues on which paleontologists are now concentrating in their efforts to unravel the events which marked the end of the Cretaceous Period. In general terms these workers can be split into two groups: those who could be called catastrophists, who favor a "big bang" extinction event; and those who could be called gradualists, who favor a slower and more protracted process of extinction.

CATASTROPHIC EXTINCTIONS

As we have already seen, many suggestions have been put forward in favor of a dramatic end to the dinosaurs: through flips in the magnetic field, exploding stars, or comet or meteorite impacts. Most have been dismissed as idle speculation by paleontologists because they seemed to represent simplistic attempts to link the big and dramatic dinosaurs with an equally big and dramatic ending. Then, in 1979, an announcement was made which immediately gave the catastrophists a far more serious and influential voice in the matter of dinosaur extinction.

The work which led up to the announcement began six years earlier as part of a routine paleontological sampling project. Dr Walter Alvarez of the University of California, Berkeley, was involved in collecting samples of rocks from limestone formations in northern central Italy, near the town of Gubbio. The samples were of tiny planktonic shelled organisms known as foraminiferans, or "forams". The shells of these creatures actually form the bulk of the limestone rocks.

The rocks in the Gubbio area were of Late Mesozoic age, and from both sides of the K–T boundary which marked the end of the Cretaceous Period and the beginning of the Tertiary Period. Dr Alvarez noted that the forams were very abundant in the limestones which occurred immediately beneath the K–T boundary layer; but right on the boundary, which is here represented by a thin band 1 in (2 cm) thick of reddish-grey clay, the forams vanish almost completely – just a single type being identifiable. Above the boundary the forams again become abundant.

In an attempt to measure how long it had taken for the thin clay layer at the K–T boundary to be deposited, Dr Alvarez returned to the University of California with some samples and described his problem to his father the late Professor Luis Alvarez, a Nobel prize-winning

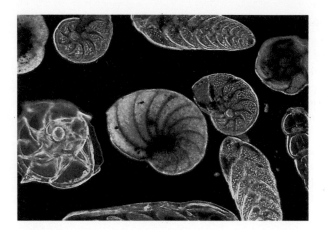

Foraminiferans live in the muddy ooze on the ocean floor. Study of their fossils inspired the Alvarez theory of mass extinctions caused by a meteorite.

astrophysicist who also worked at Berkeley. Alvarez senior devised a method for measuring the length of time it had taken for the clay to be deposited, which relied upon the steady rain of cosmic particles which falls to earth and is incorporated into any sediment. Extraterrestrial material has a different chemical signature to that of earthbound sediments, and this difference can be measured. Alvarez chose the element iridium, a metal of the same group as platinum, because it is several thousand times more abundant in meteoritic dust than it is on the surface of the Earth. Most of the iridium which would have been freely available on the Earth's surface when it first formed 4.5 billion years ago is now concentrated in the core of the Earth.

He had the K–T boundary clay analyzed for iridium, as well as limestone sediments from above and below the clay layer. The results were surprising. The clay sample was found to contain unusually high concentrations, over thirty times higher than in the limestone. This clearly showed that something of extraterrestrial origin had affected the Earth at the time of the extinction of the forams and the dinosaurs. In 1979 the team from Berkeley announced their findings, proposing that the enrichment of iridium in this clay layer may have been the result of a supernova exploding quite near our solar system, about half a light year – 6 million million miles (9.5 million million km), which is not much in cosmic terms.

This theory did not last long. Further analysis of samples of clay by Drs Asaro and Michels at Berkeley proved that the supernova theory was unlikely, because the other products of a supernova such as plutonium should also have been detectable in the sediments, and they were not. Similar anomalous concentrations of iridium also began to be detected at K–T boundary clays elsewhere in the world, first in Denmark, then in New Zealand, and almost fifty verified anomalies have since been recorded worldwide.

To explain the apparently worldwide nature of this phenomenon Luis and Walter Alvarez proposed that the iridium layer was caused by the impact and subsequent vaporization of a large meteorite approximately 6 miles (10 km) wide. The collision of such a meteorite, travelling at perhaps 60,000 mph (100,000 km/h), would have had a catastrophic effect on the Earth and its atmosphere. The impact would have resulted in the meteorite punching a hole right through the atmosphere and into the Earth's crust, where it would have vaporized and thrown up an enormous cloud of dust and water droplets. The violent ejection of dust from the impact explosion would have generated immense turbulence in the atmosphere and created winds which would have carried a shroud of dust and cloud around the entire globe. In addition to the dust and cloud cover, there would have been violent earthquakes and gigantic tsunami (tidal waves) sweeping around the world. Some support for the effect of dust clouds in the atmosphere could be gleaned from historical records of the atmospheric effects of large volcanic explosions, such as that of the Indonesian volcano of Tambora in 1816, the dust from which caused frosts in June in North America and Europe; and more recently at Mount St Helens where dust particles spread widely

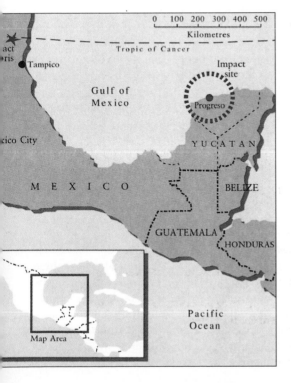

Impact site? Over the past two years a team of geologists led by Walter Alvarez has found increasingly convincing evidence for a huge meteorite impact site in the area of the Yucatan Peninsula (Mexico). Debris from the impact has been found on the mainland of Mexico, on the other side of the Gulf of Mexico. This debris includes layers of rock-droplets up to one foot thick and fossil tree remains mixed up with ocean floor sediments. The tree debris mixed up with ocean floor indicates that huge tsunamis (tidal waves) would have washed back and forth over the area after the impact, uprooting trees and hurling them out to sea.

223

around the world in the stratosphere – yet such violent explosions would have been tiny squibs compared with a huge meteorite impact.

This nightmarish event would have been disastrous for organisms living on Earth. The dust and cloud cover would have cut out sunlight for a long time – some have suggested that darkness may have persisted for months or even years. In the absence of sunlight all plants would die, destroying the food chains both on land and in the sea. The global nature of this catastrophe would seem far better to explain the very broad nature of the extinction of so many creatures at the close of the Cretaceous. Evidently the survivors would have been those best able to cope with a period of cold and dark. Plants are able to spread themselves and regrow from seeds, spores, or rootstock. Many of the animals which survived would have been small, adept scavengers such as the mammals of the time.

The scientific community reacted much more favorably to this theory than to most previous extraterrestrial hypotheses. There were two reasons for this. First, the theory had a solid foundation in observed data, which could be analyzed and form the subject of much useful debate. Second, the scenario proved to be very topical because it was relevant to military and environmental matters which have become of prime concern to all people. The similarity of the effects to those of a nuclear war – the "nuclear winter" – raised awareness of the potential of these weapons for global destruction.

Further support for the meteorite theory came from other studies of the K–T clays from various parts of the world. Droplets of rock which must have been molten and then flung out at the time of impact were found in some deposits. Also grains of quartz (sand) have been identified in these layers which, when viewed microscopically, show curiously criss-crossed fractured layers. This type of quartz is known as "shocked quartz" and is seemingly only formed under the most intensely violent conditions – such as the impact of a meteorite.

Discussion and criticism of the theory has centered on a number of issues. Most obvious among these has been the whereabouts of the impact crater left by the meteorite when it fell to Earth. Calculations suggest that the crater should have been in the order of 130 miles (200 km) in diameter. No such crater of the correct age has so far been identified, even with the aid of satellite imagery. However since approximately four fifths of the surface of the Earth is water, it seems highly probable that the meteorite hit oceanic crust, rather than a continent. If such was the case, the crater may have been lost along one of the tectonic plate margins, since about half the ocean floor which existed in the Cretaceous has now vanished into trenches along the margins of tectonic plates (see page 34).

Great volcanic activity at the end of the Cretaceous Period may have been a cause of dinosaur extinction. The Deccan of India shows that enormous volumes of volcanic lava poured out 66 million years ago. The effect of this would have been very damaging to the environment.

There has also been much debate about the length of time during which the Earth would have been plunged into darkness following the collision. The first estimates of two or three years have been rapidly revised downwards, first because such lengths of darkness would have killed absolutely everything on the planet, and second because atmospheric processes would have been capable of clearing much of the debris relatively quickly. Most scientists now favour a period of intense darkness not exceeding three months.

There has also been some debate about the significance of the iridium anomaly and its claimed extraterrestrial origin. Dr Michael Rampino of the US National Aeronautics and Space Administration (NASA) has claimed that the anomaly does not have an extraterrestrial origin at all. Rather, he says, it reflects a sedimentary mechanism – that is one involving deposition of matter on the sea floor – which led to the concentration of iridium in these clays. It is known that

225

there are anomalously high levels of iridium in manganese nodules which can be collected from the ocean floor.

An alternative to some form of sedimentary concentrating mechanism is the suggestion by Drs Charles Officer and Charles Drake that the high levels of iridium have a volcanic origin. Their studies indicate that iridium concentrations can be found above and below the levels of the K–T boundary clay, suggesting that the event was not one which was concentrated at a particular moment. There was considerable volcanic activity at this time, as revealed by the enormous volumes of volcanic lava which were deposited at the very end of the Cretaceous on the Deccan region of India, and this may also have had a considerable environmental impact. The outgassings of many thousands of volcanoes would have generated high levels of acid rain, depletion of the ozone layer, and enhancement of the "greenhouse effect" – though this last would have been counteracted by the veil of dust in the atmosphere. In addition to all these atmospheric effects, the Late Cretaceous was also a time of widespread sea regression – the seas retreated from many areas of land which they had covered shallowly during the early parts of the Late Cretaceous. The sum total of all these effects is essentially the same as with the meteorite theory, the difference being that the extinction had a terrestrial rather than extraterrestrial origin.

Clearly the debate on the proposals put forward by Alvarez father and son is still alive. Their hypothesis is supported by a considerable body of scientists – but, as we shall see, not by all. The meteorite theory as it presently stands takes it as proven that there was a meteorite impact, and that it triggered the extinction event. The main point of debate is the precise mechanism of the extinction. Was it simply the effect of prolonged darkness? Was there environmental heating through the "greenhouse effect," or cooling through the shielding effect of dust in the atmosphere? Was it the impact and the resultant fireball which produced torrents of acid rain? Was it a combination of several of the above factors?

GRADUAL EXTINCTIONS

The proposals put forward by the Berkeley team provided a great challenge to other paleontologists who favored more gradual processes for the extinction of the dinosaurs and other groups of organisms. The focus of their activities became the rocks of the very late Cretaceous, and the fine details of the fossil record that they revealed.

While Walter Alvarez was collecting samples in Italy during the 1970s, other detailed sampling was being done in the United States. This involved collecting and recording of the fauna and flora of the Late Cretaceous by Professor Leigh Van Valen of the University of

Chicago and Dr Robert Sloan of the University of Minnesota. The results of their work suggested that the dinosaurs, among others, had not passed into extinction in a single cataclysm, but had been the victims of a progressive climatic deterioration which they were able to chart across the final 5 to 10 million years of the Cretaceous Period. In essence their work revealed that during the Campanian stage of the Cretaceous (roughly between 10 and 5 million years before the K–T boundary) dinosaur faunas had been rich and diverse, and that they lived in a rich, sub-equatorial, forested environment. During the Maastrichtian stage, covering the 5 million years or so running up to the K–T boundary, faunas and floras changed. The record shows dinosaurs becoming considerably less abundant and varied, while there is a clearly burgeoning mammal fauna – known as the *Protungulatum* fauna because it includes the ancestors of the ungulates, mammals with hooves. The vegetation also changes, being more indicative of prevailing seasonal warm-temperature conditions, rather than lush and subtropical. These observations also led Van Valen and Sloan to suggest that the extinction of dinosaurs may have happened earlier in northerly latitudes than in southerly ones.

This work was practically eclipsed by the interest generated by the meteorite extinction theory, but groups of paleontologists began to explore the possibilities suggested by the work of Van Valen and Sloan in later years. In particular Professor Bill Clemens, a paleontologist based, like Alvarez, at Berkeley, began systematically collecting from the latest Cretaceous rocks along the line of the Rocky Mountains in order to check on Alvarez's theory. He and his team of collectors were able to identify an iridium anomaly, but this did not coincide with the apparent time of extinction of the dinosaurs. Dinosaurs remains always seemed to vanish from the rocks several meters below the iridium-rich layer, suggesting that the timing must have been wrong – that the dinosaurs had anticipated the meteorite extinction by several thousand years! The work of another group came up with rather different conclusions. But even they proposed a more gradual timing for dinosaur extinctions.

Did dinosaurs survive longer?

Not only Professor Clemens' results, but also the work of various paleontologists including Van Valen, Sloan, Dr Keith Rigby of Notre Dame University, Indiana, and Dr Diane Gabriel of Milwaukee Public Museum, as well as additional work involving the collection of large numbers of early mammals by Dr David Archibald of San Diego State University, all indicated that the extinction event was not instantaneous – in the sense that it took place in a matter of a few months – but rather that it lasted from 20,000 to 50,000 years. This period, which is short by geological reckoning but long in terms of animal population dynamics, was marked by an ecological succession in which communities dominated by dinosaurs were replaced by mammal-dominated communities. The key factor which seems to underlie this transition appears to be climatic – a move from balmy subtropical or tropical conditions to a more seasonal climate.

Not only that, but the collections of these researchers in the Hell Creek area of Montana seemed to show that rare dinosaurs persisted into the earliest part of the Paleocene geological epoch – that is to say, just above the K–T boundary. There is still some discussion about whether these rare remains of dinosaurs were really dated correctly. It is just possible that they may have been fossils which had eroded out of Late Cretaceous beds and had been redeposited in a Paleocene context.

Quite what caused the climatic deterioration favored by all these workers as the factor which underlay the cumulative extinctions at the close of the Cretaceous is not agreed upon. However, in general terms it is known that the closing phase of the Late Cretaceous was a time of large-scale regression of the sea, and considerable volcanic activity. The combination of these factors may have had a significant impact on climates everywhere, producing much more varied seasonal conditions as the ocean currents changed with the new shape of the oceans, and the winds changed with the currents. Global temperatures would have fallen by many degrees, creating much less favorable conditions for dinosaurs, as well as for other animals adapted to warm conditions. Isolation of faunas on smaller land areas following the breakup of Pangaea may have caused ecological pressures which less adaptable forms – which may have included the dinosaurs – were unable to sustain, and this may have contributed to their waning and final extinction.

Could both theories be right?

For the moment, the views of paleontologists concerning the fall of the dinosaurs, and of the many other groups which disappeared around 66 million years ago, have polarized into two factions, one favoring a catastrophic end, the other a gradual mode of extinction. But these two views are not irreconcilable. There are a number of people who favor the gradual model of extinction and believe that fossil collections show that many groups of animals including the dinosaurs were steadily declining in abundance and diversity toward the close of the Cretaceous. However, in the face of the impressive data which Alvarez and colleagues have assembled, they are quite prepared to accept that a meteorite impact occurred coincidentally, and perhaps hastened the inevitable end of at least some of these groups.

WAS DARWIN WRONG?

The divide among paleontologists into those who are prepared to accept catastrophic changes in the history of life, and those who advocate more gradual change, represents two differing philosophical attitudes towards the nature and history of life. These attitudes were held by two notable men whom we have already met: Charles Darwin, who favored slow, almost imperceptible change in the evolution of life; and Thomas Henry Huxley, oddly enough one of Darwin's staunchest supporters in the battle over his theory of evolution, and who was yet quite happy at the thought that evolution might progress in a much less regular manner.

Evolution by jumps

The Darwinian view of the history of life being one of slow gradual change dominated biological thought from the early decades of this century through to the 1970s. Since that time opinion, even if it has not gone the other way, has at least wavered. Doubts about the strict Darwinian view have arisen through the work of Professor Stephen Jay Gould of Harvard University, as well as several others. They have taken the view that much of the history of life can be charted as prolonged periods of little evolutionary change punctuated by periods of fast and dramatic change. This view is summed up by the phrase "punctuated equilibrium" – life progresses steadily in an equilibrium phase, but is periodically punctuated by times of rapid change.

As we have seen in the discussion of the extinctions at the close of the Cretaceous Period, our perceptions of what was happening at the time depend crucially on the quality of the record of fossils in the rocks. The Darwinian view would be that the rock record is such a small sample of organisms living at any one time, and is so liable to interruptions in deposition, that it is bound to be full of gaps, and should not be expected to show a smooth, steady evolution of life forms. This view was (and among biologists today still is) widely held to be true. However, Gould and his colleague Niles Eldredge challenged this view in the early 1970s. They looked at the fossil record, particularly of shelly marine organisms, and found a consistent pattern of rock sequences showing fossils with a very similar form existing for considerable periods, and then ending abruptly. The sudden disappearance of one such form then resulted in the equally sudden appearance of a new and different fossil species replacing the ones which had vanished. The new species would in turn continue with little change until it too suddenly vanished. The repetitive nature of this pattern led Eldredge and Gould to suggest that rather than being a chance effect of fossilization, these formations were really an indication of the mode of evolution in the history of life: periods of little change punctuated by periods of rapid change.

This challenge to the views of modern evolutionary biologists was the first of a number of incursions which paleontologists have been able to make into what has always been considered a realm where biologists working on living species are the sole arbiters. Work following up on the new and controversial ideas of Eldredge and Gould focused on the quality and detail of evidence that could be provided by the fossil record, and gave rise to a number of detailed surveys.

One of the most ambitious and thorough surveys was carried out by Dr John Sepkoski of the University of Chicago. His brief was to catalog the time of first appearance and eventual disappearance of groups (families) of marine organisms throughout the Phanerozoic Eon – there are well over 3,000 known families, comprising about a quarter of a million species. He confined his catalogue to marine organisms for the simple reason that their fossil record is more complete than that of land-living forms, and the dating of marine rocks is generally more accurate.

The main reason for carrying out this survey of the history of marine life was to see if it was possible to trace patterns of changes in diversity in the history of the groups as a whole. This aim succeeded, providing some evidence that diversity patterns, and hence the evolutionary histories of the groups as a whole, had been "reset" at intervals through the Phanerozoic by fairly abrupt dips in the variety of families – mass extinction events.

In collaboration with Professor David Raup, also at Chicago, an even more detailed analysis of the fossil record of marine organisms from the Late Permian through to the present day was assembled. This covered a timespan of 250 million years. The record seemed to Raup and Sepkoski to show a curiously regular repetition of extinctions, which fell into a cycle of approximately 26 million years. The cycle started with a massive extinction at the end of the Permian Period, which resulted in the extermination of about 90 per cent of all marine organisms. Twenty-six million years later, at the end of the Triassic, there was another extinction event; and this was repeated a further seven times up until the last, which in their estimate occurred about 10 mya. Admittedly the extinction events vary quite considerably in their intensity: three were what we might classify as major catastrophes, the rest were apparently much less pronounced.

When their results were first announced, they naturally created great interest. Numerous statistical analyses of the figures seemed to confirm their observations to the satisfaction of many. The question which remained, however, was: What could possibly have a cycle of 26 million years?

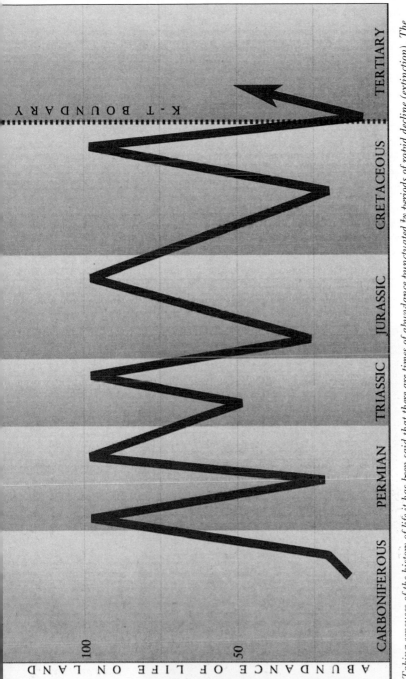

Taking censuses of the history of life it has been said that there are times of abundance punctuated by periods of rapid decline (extinction). The dip at the end of the Cretaceous is just one of several that are known to have occurred.

The death star

Astrophysicists promptly responded to this problem, coming up with a variety of possible but rather speculative mechanisms – such as periodic solar flares, or "wobbling" of our galaxy. None of these was taken particularly seriously, the main difficulty being how to prove a regular 26 million year cycle for such events. The problem also naturally came to the attention of Professor Luis Alvarez at Berkeley. Alvarez appears to have dismissed Sepkoski and Raup's model of regular extinctions. However, a former student of Alvarez's, Dr Richard Muller, was not so dismissive and the two scientists continued to argue over the subject for many months. This eventually resulted in Muller suggesting a new idea which proved to be quite feasible. He proposed that the Sun had a sister star. Many of the stars in the galaxy are paired, so it is not impossible to believe that our own Sun has its pair. He proposed that this sister star and our Sun orbited each other with a 26 million year cycle, and that once in every cycle the star passed close by our solar system in such a way that its gravitational field disturbed asteroids in the asteroid belt which lies between the planets Mars and Jupiter. The asteroids, thrown out of their former

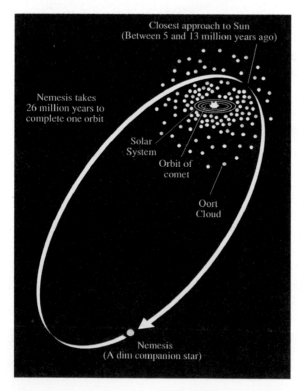

Closest approach to Sun
(Between 5 and 13 million years ago)

Nemesis takes
26 million years to
complete one orbit

Solar
System

Orbit of
comet

Oort
Cloud

Nemesis
(A dim companion star)

THE NEMESIS THEORY

The Death Star, a dim companion to our own Sun, moves in an elliptical orbit around our solar system. As it approaches our Sun it passes through the Oort Cloud, hurling comets and other cosmic debris toward Earth. If Nemesis orbits every 26 million years in this way then regular mass extinction on Earth would be readily explained.

orbits, could have showered the Solar System, and the Earth would receive its share of meteorites.

Unfortunately there were two problems with the theory. First, no companion star to our Sun has ever been observed; and second, a 26 million year orbit time could not be reconciled with an orbital path for the companion star which would affect the asteroid belt. A solution was eventually suggested by Dr Piet Hut of Princeton, who proposed that Muller's companion star did not disturb the asteroid belt during its orbit around the Sun, but in fact crossed the Oort Cloud of interstellar debris which is believed to lie beyond the orbit of the planet Pluto, and which is thought to be a source of comets. In this model the projected orbit of the companion star proved stable on a 26 million year cycle. Disturbance of the Oort Cloud could dislodge a shower of comets into the Solar System. In 1984 Muller and colleagues formally announced this theory, and proposed that the as yet unidentified companion star should be named Nemesis after the goddess in Greek mythology who relentlessly persecutes the rich and proud.

Naturally various objections have been raised to the Nemesis theory. The supposed star has continued to elude astronomers – though Nemesis may be a small, dim "brown dwarf" star, and therefore difficult to see. Also there is much concern that the orbit of Nemesis would have to be so wide that its gravitational attraction to the sun would be very weak and it could simply drift off across the galaxy, rather than returning to make a close approach.

MORE THEORIES TO COME?

The debate is clearly not yet at an end. The search for Nemesis continues unabated, and new theories will undoubtedly emerge which may draw on new facts or observations, or emerge from reevaluation of what is already known.

Despite all the problems and difficulties raised by the extinction of the dinosaurs and the other creatures that perished with them, paleontology has proved to be a rich and fertile area for ideas and theories. These, as we have seen, have implications not merely for the fate of dinosaurs or ammonites or chalky plankton but, more importantly, have led to new interpretations of the history of life.

ABOVE: *This picture provides one of the most up-to-date illustrations of* Iguanodon, *showing its alternative postures. Large adults moved mostly on four legs, younger and smaller ones on two.*
RIGHT: *A* Diplodocus *surveys the late Jurassic landscape of North America. The surprisingly small head has nostrils located on top, almost between the eyes.*

Chapter Eight

A Personal View

Much of what I have written thus far seems to me to be what you would expect from a scientist. The tone is impersonal, suggesting a disinterested observer looking down upon the debate and selecting this argument or that, apparently with infinite wisdom. Of course any wisdom I may have is that of hindsight – being able to look back across the years at the arguments and how they have developed, and to pick and choose those that seem to have survived the best. The reader will not so far have been given much of an impression of what or how I think as a paleontologist about dinosaurs and the debates which surround these perplexing creatures.

To make amends for this I should like to look at a little of the research with which I have been involved over the past few years. And I should also like briefly to summarize how I think dinosaurs lived, and why I think they were so successful during the Mesozoic Era.

My introduction to dinosaurs came in the form of *Iguanodon*, one of the first dinosaurs to be studied. It was first named and described by Gideon Mantell in 1825, and in 1841 became one of the three founder members of Professor Owen's Dinosauria. Since then it has been the subject of a considerable amount of research by British and European paleontologists. Louis Dollo devoted a considerable part of his career to the study of large numbers of *Iguanodon* recovered from a coal mine in Belgium. In view of all this it would seem to be an exceedingly poor choice of animal to embark upon, especially at the beginning of one's career. Surely, so the argument goes (and it was an argument I heard frequently when I started out on this work), all the important work on this animal has already been done – why not look at something new and unexamined? Such a project would be much more likely to yield new and exciting results.

There is truth in this, of course, but – and there has to be a "but" – sometimes looking at animals anew can provide unexpected results. Ideas, attitudes, and techniques change over the years and these can often generate new ideas, observations, or theories, especially when they are applied to "old" material. Paleontology can be studied in more than one way; and over the past few decades this is being appreciated more and more. One line of study is exemplified by the work of Jack Horner, which has brought to light completely new and very exciting discoveries, such as the nest sites and embryonic remains of dinosaurs, during field work in Montana. These discoveries have generated new data and new theories relating to the life and habits of dinosaurs that were previously unimaginable. Another line of study is far less "glamorous" than finding new specimens in the field; it involves patiently poring over museum collections and searching through dusty museum drawers, among the broken bones and fragments. That is just where John Ostrom found some vital clues to the dinosaur origin of birds. Yet another style of investigation is to restudy specimens after many years have elapsed and techniques of interpretation have changed; this is really a form of checking up on old ideas, but often produces surprises. Finally, of course, there is a "synthetic" approach which seeks to use elements of all these methods to get as near to a truly balanced view as possible. All these courses are being pursued today by researchers in museums and universities around the world, and the contribution of each is vital to the healthy growth of the subject.

How Iguanodon *stood and walked*

The theories on the posture of this animal have a 150-year history. Not only does this make a good story; it is also important, because it charts the evolution and growth of scientific thought concerning the structure, and to some extent the relationships, of dinosaurs.

During the earliest work on this animal Mantell, with some help and guidance from Cuvier, eventually came to believe that the teeth which he had discovered belonged to an extinct lizard related to living iguanas – but 60 ft (18 m) long. He brought this view to some form of reality in the 1830s by means of an ink sketch of the restored animal (see page 71).

Mantell's lizard-like version of *Iguanodon* did not last. In 1841 Owen provided a radically different model for this animal as a dinosaur – even though it was based upon little more evidence than was available to Mantell. It was, he estimated, no more than 30 ft (9 m) long; had four large, sturdy legs, and a short neck and tail; and resembled a gigantic scaly rhinoceros – as seen to this day in the Crystal Palace models created under Owen's supervision in 1854.

Dinosaurs were not lizards, as Mantell and others had thought. They were exotic reptiles, superior to any modern species and much more akin in their bodily construction to the large tropical mammals of today. Given the fossil materials available at the time Owen was remarkably far-sighted in this proposal, but his model was not to last very long either. Opinions began to waver in the late 1850s about the correctness of Owen's models – even his own research pointed to a number of inconsistencies. The Owenian dinosaurs were ruthlessly and somewhat unfairly overturned by Thomas Huxley in the late 1860s in favour of a much more bird-like model.

Huxley was able to point to a number of bird-like features in the anatomy of *Iguanodon*, and recommended a radically different body form, much like that proposed for *Hadrosaurus* by Joseph Leidy. In 1878 Huxley's theoretical arguments became factual ones with the discovery of complete skeletons of *Iguanodon* in the coal mine at Bernissart in Belgium. Dollo's work paid clear tribute to the work of Mantell, Owen, Leidy and Huxley in gradually unveiling the true nature of this dinosaur. The body proportions of entire skeletons of the dinosaur were surprisingly close to those of kangaroos, supporting the views of Leidy, and the bird-like character of the hips, legs and feet was unmistakably as Huxley had predicted. Of course subtle changes and surprises emerged from a minute study of complete skeletons, such as the clear demonstration that the conical horn placed upon the nose by Mantell was in fact a much enlarged thumb claw.

Dollo's reconstruction of *Iguanodon* is one that has been followed almost universally since the early 1880s. Standing some 15 ft (4.5 m) high, this dinosaur is almost invariably pictured browsing from trees, using its tail rather like a shooting stick in order to rest comfortably while reaching into high branches. Dollo's researches had suggested that the backbone was constructed to support the animal in this posture, and that was at least part of the function of the bundles of bony rods which had been found running along it. In general terms Dollo considered this dinosaur to be the broad ecological equivalent of the modern giraffe. He even went so far as to discover evidence for a prehensile tongue – further confirming the giraffe-like nature of this dinosaur.

Despite Dollo's preeminent position with respect to *Iguanodon* Gerhard Heilmann, who wrote that important book *The Origin of Birds* (page 191), produced a very dynamic reconstruction of *Iguanodon*.

However, this interpretation seems to have been overturned by Dollo's own work. In 1928 Heilmann produced a "new restoration of *Iguanodon*" which was far more upright, walking with its tail dragging on the ground, in a short paper in honour of the work of Louis Dollo. Bearing in mind that Dollo had complete skeletons to work with, it is not really surprising to find that his particular reconstruction of this dinosaur became universally accepted. However, further examination of this material has revealed a number of surprising details.

The tail of *Iguanodon*, with its upward sweep, provides the starting point for a new look at the posture of this dinosaur. All the skeletons discovered in the mine at Bernissart – several of which are laid out in the positions in which they were found, on banks of plaster in the museum in Brussels – seem to have tails which are held either straight, or dipping slightly downwards at their ends. Indeed they would have been held more or less in that position in life by the arrangement of bony tendons lying alongside the spines of the vertebrae. Careful examination of some of the mounted skeletons shows that the tail has been broken in order to create the upward sweep! Dollo was clearly making his dinosaur fit the prevailing views of Huxley and Leidy by giving it a kangaroo-like tail.

Looking at the skeleton in simple mechanical terms, straightening the tail tilts the animal forward and downward, so that it becomes balanced at the hips. This arrangement makes great sense because it allows the network of bony tendons, which are distributed equally in front of and behind the hips, to support the backbone in tension most efficiently. Lowering the chest towards the ground also brings into question the function of the arms and hands. The arms are long and comparatively strong, and the shoulders and chest are also very heavy-boned. That suggests that the arms could have been used for more powerful movements than simply grasping twigs and branches.

Examination of the hand seems to confirm this very strongly. The wrist bones, which in other bipedal dinosaurs are small and rounded so that the hand can move freely on the end of the arm, are here heavy, block-shaped, and welded together by strands of bony tissue. The strands would appear to be sheets of ligament which have turned to bone in order to strengthen the wrist, and reduce the mobility of the hand.

The bones of the hand are surprisingly modified. The first finger, the equivalent of our thumb, has a very short palm bone (metacarpal) which is actually welded into the wrist. The remainder of the finger consists of two bones. The first is a small, flat plate, which slides on the metacarpal and fits into a pit on the inner side of the top joint. This joint is cone-shaped and enormously large, and

RIGHT: *Hawkins' model for the Crystal Palace* Iguanodon *may look antiquated in out eyes now, but at the time this bear-like creature was the very latest interpretation that science could offer.*

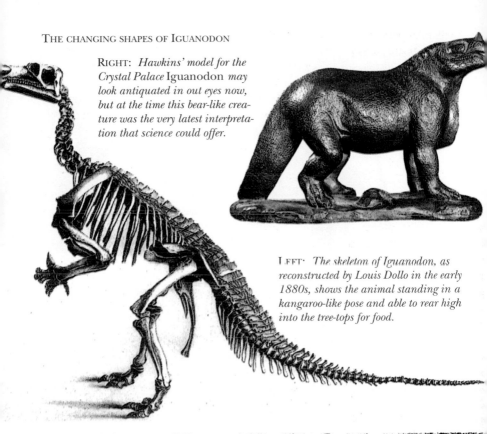

LEFT: *The skeleton of* Iguanodon, *as reconstructed by Louis Dollo in the early 1880s, shows the animal standing in a kangaroo-like pose and able to rear high into the tree-tops for food.*

RIGHT: *Heilmann's 1916 drawing of two* Iguanodon *moving with some alacrity contrasted very strongly with the slower and more ponderous restoration that he produced about 10 years later.*

How the hand worked

The flexible fifth finger moves a bit like a human thumb for grasping objects, while the middle three fingers are capable of little flexure.

The large, stiletto-like thumb spike of Iguanodon *would have been a devastating weapon. The sharp spike, coupled with the strength of the forelimb, could have punctured the toughest skin.*

For walking, the hand acts as a foot, the middle three fingers are splayed out and end in flattened hooves.

was evidently covered in life by a very long, sharply pointed claw. The entire digit sticks out from the wrist at right angles, pointing away from the other fingers of the hand. It was clearly not used for holding, and seems most likely to have been a vicious stiletto-like spike for close combat with other dinosaurs.

The second, third and fourth fingers are more normally proportioned. Each has a long metacarpal, while the finger bones are short and block-shaped, ending in unexpectedly broad, flattened hoof-shaped claws. These finger bones are also unusual because, rather than folding inwards so that they can be used for grasping and picking up objects as a true hand normally might be expected to do, they have joints which allow them to be bent backwards and spread apart in just the same way as the toes of the foot.

The fifth finger is again different from the rest. The metacarpal diverges from the others and is short but highly mobile, being able to move very freely against the wrist; rather like the thumb spike, the whole finger diverges sharply from the others of the hand. The end bones of the finger taper to a slender tip and also seem to be highly mobile. The arrangement of the bones in the fifth finger is much as would be expected in a typical finger of a grasping hand, combining slenderness with flexibility, and lacking a large claw.

In fact, *Iguanodon* had an incredibly specialized hand, far more so than anything else that I know of in the animal kingdom. It had a thumb which was used as a defensive weapon. The three fingers in the centre of the hand were specialized as walking toes – why else should they end in hooves? This function also explains the great strengthening of the wrist bones. The fifth finger was used for grasping in the normal way.

These facts about the hand, combined with the observations on the spine, suggest a posture and method of locomotion for *Iguanodon* quite different from that provided by Dollo. The backbone would normally have been held more or less horizontally, and the hand would have been used for walking upon – though clearly the bulk of the body weight was carried by the back legs. This is clearly a much less bird or kangaroo style of posture. However, the specialization of the thumb as a defensive spike, and of the fifth finger for grasping, clearly point to the animal's ability to rear up on its hind legs alone. Height can be an advantage for fighting predators and would also have enabled these animals, as Dollo correctly surmised, to reach into trees. The prehensile finger may well have been used to grasp branches or strip them of foliage.

There is no doubt that *Iguanodon* walked on four feet for at least part of the time, but some additional evidence points to a possible change in style of locomotion with growth in these dinosaurs. Smaller individuals appear to have proportionately shorter arms than fully grown individuals. This suggests that young dinosaurs may have spent much more of their time running on back legs alone. If this was the case, it would seem to make some biological sense. Younger, smaller individuals would have been relatively more vulnerable to predators than fully grown, powerful adults. Thus an ability to be more nimble and run faster could have been advantageous for survival.

Muscles, brain, nerves, and blood

In addition to the simple mechanical shape and arrangement of bones which can lead to a detailed understanding of the shape and posture of the animal as a whole, individual bones may also leave clues about the soft tissues of the animal. This is particularly the case where powerful muscles are concerned. Muscles, particularly those of the limbs, need to be powerful, and must be very firmly anchored to the bones of the legs. They are attached either by tough sheets of connective tissue or by cord-like tendons. Where the sheets or tendons are attached the bone is frequently roughened, or in some cases develops into crests, ridges, or bony spurs.

Detailed examination of the surface of the bones of the legs, shoulders, and hips reveals a pattern of "muscle scars" on the bones indicating where muscles used to lie, and in some cases how powerful they were. It is clearly never going to be possible to reconstruct all the muscles of an extinct animal from such evidence, because many muscles do not leave such clear evidence of their attachment points. Nevertheless it is possible to build up an idea of the distribution of some of the more important limb-moving muscles. In addition to the

Reconstruction of Iguanodon's main hip and thigh muscles.

information on muscle scars, the most important clues come from muscle arrangements in living relatives of the fossil animals, in this case birds and crocodiles.

The result of this work can be in the form of technical drawings indicating the directions and arrangements of muscles in some of the more important areas of the body, such as the hips, shoulders, and head. Other areas of the body do not show such clear evidence of muscle attachment and can therefore be reconstructed much less precisely.

On even rarer occasions it may prove possible to learn something of even softer tissues than muscles. For example around and within the head there are many spaces in the bone which either act as containers of soft tissue (such as the brain case and eye sockets) or as passages for nerves and blood vessels. Once any creature has been dead and buried, these soft tissues rot away quickly and the spaces are filled with sediment. Once the sediment and bone are fossilized it is often difficult to separate them, which makes it hard to discover the shape of the cavities. The bones may also be crushed or distorted, in which case it is impossible.

Occasionally, however, specimens are found in which the tissue spaces of the head are filled with a clearly distinguishable sediment, forming a "natural cast" of their shape. Just such a specimen has been discovered of *Iguanodon*. It was one of those finds in museum drawers, in this case those of the Natural History Museum in London. It consisted of the back part of a large skull, the bones of which were in very poor condition. On examining this I realized that it was just possible that the brain case was still intact and filled with very hard siltstone. I had the specimen cut in half with a diamond saw. It became clear that there was a natural cast of the brain cavity inside. Peter Whybrow of the museum laboratory spent several months carefully removing the bone of the brain case to expose the stone-filled cavity within, and the results were spectacular.

The specimen showed the shape of the cavity in which the brain sat, the passages for all the nerves running to and from the brain (including those going to the nose, eyes, ears, mouth, and tongue); the passages and chambers taking blood to and from the brain; the detailed structure of the inner ear, including the hearing region (lagena) and balancing organs (osseous labyrinth). In the floor of the braincase, tucked just beneath the brain, it was also possible to see the shape of the pituitary gland.

This fortunate discovery has made it possible to learn something of the size and shape of the brain of this dinosaur, its sensitivity to different stimuli, the acuteness of its ear, and its circulatory system.

THE *IGUANODON'S* BRAIN: DECIPHERING THE EVIDENCE
RIGHT: *This diagram picks out the details of the cast, identifying the positions of soft tissues, veins, nerve passages and sense organs. From such information it is possible to build up a significant picture of how the brain, and therefore the animal, functioned.*
LEFT: *For the paleontologist, this unpromising lump of fossil provides detailed information about the structure of* Iguanodon's *brain. It is a natural cast of the skull cavity in which the brain sat.*

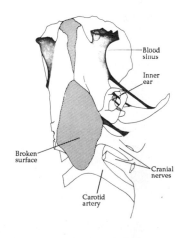

Blood sinus

Inner ear

Broken surface

Cranial nerves

Carotid artery

How Iguanodon *fed*

Jaw muscles are powerful and leave distinctive scars upon the bones of the jaws and skull, which allow the pattern of muscles in the head to be reconstructed. But a more useful guide to a creature's feeding habits is the shape and arrangement of the jaws and teeth, and the way in which the teeth are worn. These are very clear mechanical indicators of the way the jaws moved. Carnivores tend to have very similar jaws and their teeth are a mixture of stabbing and slicing blades, since meat is of a similar consistency no matter what animal it comes from. Herbivores have a more difficult time. Plant food is very tough and requires a "food processor" to break it down into a digestible form; this can be either in the mouth or in the stomach depending on the strategy employed by that particular animal.

Iguanodon had a large, rather horse-shaped head equipped with many teeth – it was clearly a mouth grinder, rather than a stomach grinder. Large size implies a substantial diet, and this in turn constrains the animal to be as efficient as possible in the way that it manages its food.

As with all ornithischian dinosaurs there are no teeth at the tips of the jaws. *Iguanodon* has a broad, sharp-edged beak; in life the bone would have been covered by a self-sharpening horny outer beak, similar to that of a tortoise. The front edge of the beak was irregular, and this probably improved its cutting abilities. This arrangement is ideal for an intensive plant feeder because not only is the beak sharp and good at cutting extremely tough pieces of vegetation, but there is no danger of it being blunted during the lifetime of the animal since the horn which forms the beak grows constantly, and the outer layer is harder than the inside so that it wears at an angle and always has a chisel edge.

Plant-eating reptiles living today do not chew their food. Tortoises and various lizards (including iguanas) take simple bites of food, give

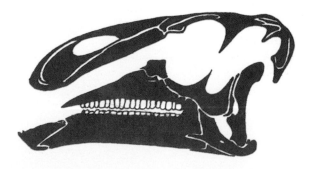

Iguanodon's *skull had unexpected hinges across the side of the face, allowing the upper jaws to move as the jaws closed.*

one or two gulps and swallow the food with practically no treatment; they rely upon their stomach to digest it slowly. Their teeth or beaks reflect this; their teeth are simple and shown no signs of special wear. Dinosaurs such as *Iguanodon* do not seem to have been able to take such a leisurely approach to the digestion of plants. Its jaws are lined with large, leaf-shaped teeth, fifty or more in each jaw, which interlock with one another to form grinding "batteries." The teeth also wear down at an angle along the grinding surface. The scratch marks on the worn surfaces of these teeth show quite clearly that these reptiles were very capable chewers.

The teeth are markedly inset from the sides of the mouth, leaving a pouch-like area outside the teeth, and against the side of the face. As Professor Richard S. Lull, and much later Professor Peter Galton of the University of Bridgeport, were able to argue, this arrangement indicates the presence of fleshy cheek pouches which are necessary to catch the food sliced between the teeth. No living reptiles are known with cheek pouches. That is a strictly mammalian trait in the modern world, but then – as I will argue later – these were no ordinary reptiles. Cheek pouches are necessary for effective chewing without dropping food.

The structure of the bones of the skull and the way that the jaws hinge against the skull are quite remarkable. A flexible hinge runs diagonally across the side of the face from just behind the beak region, up across the face, across the eye socket, to the top at the back of the head. This hinge allows the upper teeth to swing inward and outwards as the jaws open and close. As the jaws are closed the teeth in the lower jaw slide across the inside edges of the upper teeth, forcing these to swing outwards. The amount of movement of the teeth in the upper jaw was severely limited by very powerful ligaments within the skull.

This system seems extraordinarily complicated, but was essential for efficient chewing, which depends on teeth rubbing past each other. Mammals have solved the problem by developing lower jaws that can be moved sideways, but no reptile has ever been able to do this, since reptiles' jaw muscles allow only up-and-down movement.

Chewing and evolution

The demonstration that dinosaurs such as *Iguanodon* were able to chew may not seem so important at first sight. However, I think that this may well have been of very considerable importance to dinosaurs, their evolutionary history, and the history of the plants upon which they fed.

Once I had noticed the curious jaws of *Iguanodon*, I started to look elsewhere among dinosaurs to see whether others had used a similar

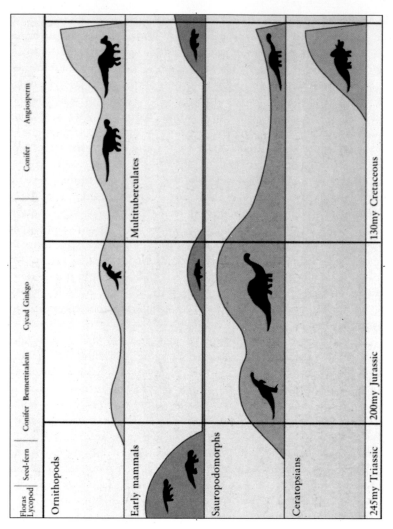

Looking across the history of the dinosaurs it is possible to see general changes in populations which may indicate large-scale evolutionary change. Dinosaurs arose in the Late Triassic and during the following Jurassic Period the sauropodomorph dinosaurs were most abundant – these were the first land animals capable of browsing on trees. In the Cretaceous Period the sauropodomorphs had declined in favour of the ornithopods and ceratopians. It may be that these changes in fortune reflected the changing abundance of plants, and in particular the rise of smaller, shrubby flowering plants (angiosperms) in the Early Cretaceous.

mechanism or whether it was unique to this dinosaur. I found it in a range of ornithopods (hypsilophodontids), various close relatives of *Iguanodon*, and the hadrosaurs or duck-billed dinosaurs. Charting the evolutionary history of the ornithopods, it is becoming increasingly clear that they started, in the Early Jurassic, as relatively rare, small forms among the general population of dinosaurs. During the later Jurassic some larger forms, such as *Camptosaurus* and *Dryosaurus*, appear still to have been relatively rare – as far as we can tell from fossil collections. However, in the Cretaceous Period there is a spectacular increase in both the size and abundance of these dinosaurs. In the Early Cretaceous there is little doubt that *Iguanodon* was one of the most abundant of herbivores. Large numbers of skeletons or partial remains have been found right across Europe (England, France, Belgium, Spain, Germany) into Asia, and more recently in various sites in North America. The species may not have been very varied, but they certainly seem to have been abundant. In the Late Cretaceous the hadrosaurs seem to represent the last fling of the ornithopods. Not only are their remains found abundantly, but they now also seem to be very varied, many different types having been identified by their distinctive headgear.

The steady increase in abundance and variety among the ornithopods through the Mesozoic seems to reflect at least partly the equally steady decline in abundance of sauropods, which were one of the other main groups of herbivorous dinosaurs. This interchange in fortunes of the two groups appears to show a coincident shift from stomach grinding (the technique of the sauropods) to mouth grinding. The Early Cretaceous is also a time of floral change, and this may be at least partly connected with the changing fortunes of these groups. The flowering plants seem to make their first appearance during the Early Cretaceous and begin to diversify and become dominant throughout much of the later Cretaceous.

What these changing patterns of diversity mean is presently uncertain. However, there is no doubt that in broad terms they exist, and reflect underlying long-term changes in various groups. It is tempting to suggest that there is some linkage between these changes. For example, sauropod stomach grinding seems to be the most successful strategy during the Jurassic – if "success" is measured in terms of the relative abundance and diversity of this type of animal among fossil dinosaurs. During the Cretaceous mouth grinding seems to be a more successful strategy, again reflected in the increased abundance of ornithopod over sauropod fossils.

The sauropods may have been so effective at feeding upon Jurassic vegetation that they provided an evolutionary opportunity for new, more rapidly reproducing plants to gain a foothold in cleared areas of land left in the wake of passing herds of sauropods. This change in plant communities may have favored the feeding technique used by ornithopods, which consequently increased. "Dinoturbation" – trampling of the ground by dinosaurs – may also have played a part (see page 187).

Alternatively, the change in plant types may have been triggered by subtle climatic changes which occurred across the Jurassic–Cretaceous boundary. The Early Cretaceous certainly seems to be a time of somewhat drier climatic conditions than the Late Jurassic, and again this may have favored some more tolerant, hardy angiosperms (flowering plants). In this case the change in dinosaur populations would be shadowing climatically induced and floral changes, rather than being responsible for them.

It is also possible that the evolutionary history of plants was influenced neither by dinosaurs nor by climates, but by its own internal dynamics. Increasing competition for space and light among plant species may have favored those able to reproduce and grow more quickly in order to colonize new areas of land.

None of these suggestions qualifies as a real theory to explain what may have been happening in a broad evolutionary sense during the Mesozoic. However, each suggestion is valuable in the sense that it provides a number of working hypotheses.

My suspicions are that the two groups have an interlinked history – that is to say, they show a coevolutionary relationship, changes in either group affecting the evolutionary history of the other. Colleagues such as Dr David Weishampel, vertebrate paleontologist at Johns Hopkins University, Baltimore, Dr Bruce Tiffney, paleobotanist at the University of California, Santa Barbara, and I are looking in detail at the food quality and structural strength of Mesozoic plants, their distribution and habitat preferences, and the corresponding details of dinosaur feeding capabilities, the power of their feeding apparatus, and their distribution patterns.

Iguanodon *as a living animal*

My work on *Iguanodon* has been an attempt to bring this dinosaur back to life – much as Louis Dollo attempted to do from the early 1880s through the 1920s. The fact that my interpretations differ from those of Dollo most definitely do not mean that I am in some way superior to him. I have had the enormous benefit of all his work to focus my thoughts. I have also profited from all the advances which have taken place in the science of paleontology since Dollo's time. Even so, compared with what we know about a living animal, the sum total of the work on *Iguanodon* may not seem very impressive.

The bony anatomy of this dinosaur is known in considerable detail. Some parts of the structure of the brain, the cranial nerves, inner ear, blood circulation and the pituitary are known, as is the muscular anatomy of the hips and upper leg, the shoulder and upper arm and the head and jaws in particular. Some skin impressions have been found, so that the texture of the body is quite well known.

A little is known of changes in body proportions with growth from the discovery of young specimens. Anatomy and some trackway evidence suggests that adults were capable of walking either on all four feet, or two hind feet alone. Younger individuals may well have moved around on their hind legs most of the time. Weighing somewhere between 1 and 2 tons as adults, these animals were surprisingly agile, probably being capable of running at speeds approaching 22 mph (35 km/h).

No eggs or nests have been found, so nothing is known of their egg-laying abilities or whether they looked after their young. There are no clear anatomical indicators such as crests, frills, or horns which would allow us to distinguish males from females.

The abundance of remains shows that these dinosaurs were very numerous during Early Cretaceous times. One apparent flash-flood deposit from Germany strongly suggests that they moved about in herds. These animals were large, highly efficient herbivores, and in general terms would have filled a similar ecological role to that of a modern horse or large antelope, which would seem to accord well with the notion of herds. The large thumb spikes indicate that they were certainly capable of defending themselves against predators.

We know absolutely nothing about the creature's body temperature, respiratory rate, heart rate, or indeed anything much about the tissues and tissue metabolism, although some information on its growth has come from studying bones.

Compared with the data on a living reptile this list is rather pitiful. Nevertheless the fact that there is a list at all is an achievement. All a paleontologist starts with is a pile of dusty bones, and it is a minor miracle that we are able to reconstruct anything at all, let alone become involved in quite detailed discussion and research on the evolution of feeding and plants, the body temperature and activity levels of dinosaurs, and many other problems.

So much for a single species. But, as this book has shown, dinosaurs are enormously varied. Is it possible to say anything meaningful about dinosaurs in general? A great many things have been claimed for them in the past, not least through the debate over whether they were endotherms or ectotherms. To close this chapter I will look at a number of themes relating to dinosaurs and their ways of life, by reference to two sharply different types of dinosaur.

DINOSAURS AS DINOSAURS

I am going to compare two dinosaurs, *Brachiosaurus* and *Deinonychus*, because they represent two extremes of dinosaur design; the former a 20 or 30 ton lumbering, plant-eating monster; the latter a 175 lb (80 kg) nimble, predatory creature. If dinosaurs do have some unity of design it should emerge through the comparison, and that might allow us to come to some clearer idea of what being a dinosaur really means. Alternatively, if they prove to be irreconcilably different it may prove necessary to adopt a more flexible approach to the concept of a dinosaur.

The structure of dinosaurs

Quite obviously the basic designs of *Brachiosaurus* and *Deinonychus* are radically different: one is a heavy quadruped, the other a lightly built biped. Yet within these differences there lie similarities of structure which have allowed for both designs. Most important among these are the upright legs found in all dinosaurs.

Vertical **legs** can do either of two things. In the case of *Brachiosaurus* they act as pillars to support the great weight of the body. All of the legs were held very nearly straight during the weight-carrying phase of the stride for the simple reason that bending the legs while they were supporting many tons would impose such stresses on the leg bones that they would break. For just the same reason elephants hold their legs very straight beneath their bodies as they walk. All such animals are referred to as *graviportal* (meaning "heavy carrying").

Alternatively legs held beneath the body can have a long stride, and can allow an animal to move very quickly – to be *cursorial* (meaning "running"), the technical name for all fast-moving animals. Such creatures tend to be light, so their legs can be slender, and do not need to act solely as pillars. The legs are much freer to bend through the stride to give added thrust. These are just the types of adaptation seen in a fast runner like *Deinonychus*.

The **feet** of *Brachiosaurus* are broad and rounded, providing a large surface area across which to spread the body weight; those of *Deinonychus* are by comparison tiny, with all the weight carried on two slender toes on each foot. The fact that it can bear the weight of its body on two toes only has allowed it to specialize one of its other toes for use as an offensive sickle claw. The lightness of the feet is also of great importance to a fast runner. In all cursorial animals the feet are small: for example, horses have a foot reduced to a single hoofed toe. The lighter the feet the faster the legs can be moved.

Nevertheless there are underlying similarities. In both animals the ankle is held clear of the ground; the animals are digitigrade (walk on their toes), another constant feature among dinosaurs. In the cursor this allows for a long foot which can be added to the length of the stride. In the graviportal animal the high ankle (or wrist) supported from beneath by a form of heel pad means that the foot does not have to be flexed and extended as the animal walks. This saves considerable amounts of muscular energy which would otherwise be wasted.

Weight distribution and support may appear to be very different in these two animals, but yet again there are common themes to be found. The massive gut necessary for a herbivore obliges *Brachiosaurus* to be quadrupedal, which means that the weight of the animal is distributed more evenly between front and back legs.

The **back** of *Brachiosaurus* is arched and held under tension by ligaments, rather like a bridge, to provide the most efficient support for the weighty belly region. Even though all *Deinonychus'* body weight is carried by the hind limbs alone, its back is still arched, and held in tension by stiff ligaments and muscles; this obviously also supported the belly region, but also provided a rigid chest which acted as a firm anchorage point for the powerful shoulder muscles.

The **tail** is long, slender, and held clear of the ground in *Deinonychus*, and acts as a cantilever to balance the front half of the body over the hips. It also serves as a dynamic stabilizer to allow the animal to swerve quickly while running. None of these functions for the tail can possibly apply to *Brachiosaurus*. In fact in engineering terms its tail would seem to be of little value. However, such a large tail is unlikely to have been retained if it was effectively useless to the animal. Footprints of these animals indicate that the tail was usually held clear of the ground, and in life it probably served at least three important functions. First, it provides anchorage for the large retractor muscles that pull the legs backwards. Secondly, it also serves as a cantilever, not for balancing the body at the hips as in *Deinonychus*, but to counterbalance the extremely long neck. The neck and tail are interconnected by long tendons which run along the top of the back bone. The tail may well also have served as part of the animal's temperature control system.

The **neck** in these dinosaurs is radically different, even though the number of actual neck bones in each is almost the same. That of *Deinonychus* is delicate and compact, with much flexibility and many ridges for attachment of powerful neck muscles which were important for wrenching movements of the head during feeding. By comparison the neck of *Brachiosaurus* is gigantic – one individual neck bone is as long as the entire neck of *Deinonychus*! However, the neck bones of *Brachiosaurus* have a surprising delicacy of form themselves, as can be appreciated if one is looked at carefully. Each bone is highly sculptured: the sides of each spine are covered by a network of pouches, and are honeycombed internally; the sides are cut away to leave large spaces; and the ribs which are attached to the sides have been found to be not only long and slender, but also hollow. Thus even though the neck of this dinosaur is spectacularly long, it is a surprisingly light and mobile structure because the individual bones are light and honeycombed. Such a structure has the greatest possible strength for its weight.

The neck would not have been thickly muscle-bound. There would have been sheets of muscle running between each vertebra so that the neck could be bent with great precision. However, larger movements of the whole neck would have been made by a more remotely controlled system. Enormously powerful elastic ligaments stretched between each of the upright spines of the bones of the neck and back, and would have kept the neck in a raised position without any muscular effort. Other long tendons running from the back up the back of the neck would have operated like the cables of a crane jib to raise and lower it. Beneath the neck on either side can be seen the long, thin ribbons of bony ribs, to which strong neck muscles would have been attached. These would have been used to bend the neck from side to side while the animal was feeding among the treetops.

Finally there are the **heads**. Clearly since one animal is a carnivore, and the other a herbivore, there are necessary differences in the jaws. *Deinonychus* has sharp, blade-like teeth, while *Brachiosaurus* has spoon-shaped teeth. But both skulls are surprisingly lightly constructed, even though that of *Brachiosaurus* is over 3 ft (1 m) long. They have large openings in the sides, again combining lightness with strength.

The physiology of dinosaurs

Clearly it is more difficult to be precise about the workings of the soft anatomy of dinosaurs. Yet it is possible to draw some basic comparisons.

We can be sure about some features of the blood circulation of *Brachiosaurus*. As discussed earlier (page 168), the height of the head above the position of the heart renders it virtually certain that the heart of this dinosaur was not only massive – in order to generate enough pressure to pump blood all the way up to the brain – but also fully divided, ensuring that blood at very high pressure did not leak into the lungs. But the difference in head and heart height can be taken further.

It is noticeable that all living reptiles standing in their natural posture have their heads almost exactly on a level with their hearts; this fits with the fact that their hearts cannot properly separate pulmonary and systemic blood. Yet *Deinonychus* and almost all dinosaurs have elevated heads in their reconstructed postures. The implication from this is that not only *Brachiosaurus* but all other dinosaurs, including *Deinonychus*, had a fully divided heart. Such circulatory efficiency greatly improved their ability to sustain high levels of activity – blood moving around the body at high pressure is able to supply the essential food and oxygen more rapidly to the muscles.

Respiration is another function which can only really be guessed at in dinosaurs. Yet again a few hints can be gleaned. The large hollows in the vertebrae of *Brachiosaurus* have already been mentioned. The ribs also have openings near their tops and hollows running through their centers. These features were of obvious value in reducing the weight of the animal. However, there is evidence from living animals that they have another function. Modern birds have hollows in their backbones and limbs and honeycombed bones; these are connected to an extensive system of air passages which form part of the respiratory system. They are connected to a system of large air sacs, which operate somewhat like a system of bellows for pumping air through the lungs. They also seem to remove heat created by the large, powerful flight muscles.

If the openings and passages in the skeleton of *Brachiosaurus* and many other of the larger saurischian dinosaurs correspond to those of living birds, dinosaurs would not only have had a highly efficient respiratory system (the system in birds is far superior to that of mammals) and perhaps have been capable of very high activity levels; they may also have used the system to dump body heat, as we shall discuss later.

Deinonychus does not show such modifications to the ribs and vertebrae, but large theropods such as tyrannosaurs do, so it is probably not unreasonable to suppose that it also had a highly efficient respiratory system. That would in any case be a prerequisite of such a fast-running, predatory animal.

Temperature control is a subject which has already been mentioned at some length earlier in the book (pages 131, 171), so I will not treat it in much detail now. In this case the comparison points to clear differences between *Brachiosaurus* and *Deinonychus*.

A stable internal body temperature – no matter what the actual temperature – is very desirable for any animal. Constancy of temperature means that the internal chemistry of the animal can continue smoothly and efficiently under all circumstances. All chemical reactions in the body are controlled by enzymes and these are very temperature-sensitive. It is no coincidence that the animals with the largest brains today – the birds and mammals – maintain a constant body temperature. The brain relies on an incredibly complex series of enzyme-controlled reactions, and such a sophisticated structure could not have evolved had body temperatures not been held constant.

Brachiosaurus, thanks to its enormous size, and hence its very high ratio of volume to surface area, would not suffer rapid changes in temperature. As Professor Colbert and colleagues was able to demonstrate with crocodiles, the larger the animal the slower its body temperature changes with changing external environmental temperature.

As a survival strategy this has a number of benefits, especially for a brachiosaur. It can generate body heat very rapidly by muscular activ-

Necks

Despite the huge difference in size, both Deinonychus *and* Brachiosaurus *have strong, highly flexible necks with approximately the same number of bones.* Deinonychus' *neck needs to be strong to anchor the head when tearing lumps of flesh from its prey.* Brachiosaurus *neck operates like a crane jib, moved by "cables" (ribs) to which powerful neck muscles and ligaments are attached.*

MUSCLES OF THE *BRACHIOSAURUS*
Using the skeleton as a starting point, it is possible to trace out the area of attachment of ligaments around the joints of the limbs, and the attachment areas of tendons of the larger leg-moving muscles. The muscle distribution can then be traced on to the skeleton and finally the skin can be added to show the animal as a living creature.

Legs
The long stride of Brachiosaurus *allowed it to cover the ground surprisingly quickly, but the legs were really designed to act as weight-supporting pillars. The legs of* Deinonychus *are long and slender and can be moved very quickly, providing great acceleration.*

Feet
Small, light feet are sufficient for a lightly built predator but not for the giant, which needs feet designed to support great weight. However, in both cases the ankle and wrists are held up off the ground.

ity – simply walking around requires a great deal of muscular work, and the heat from this is carried around the body in the blood. Add to this the fact that these animals carry around with them a very large fermentation tank in their gut, which would also generate heat, and it seems almost certain that these animals were consistently warm.

In fact they may well have had problems with overheating, rather than keeping the body warm (which is the traditional way of approaching the discussion of temperature control). Relatively mod-

Backs

The arch of the back enables the Brachiosaurus *to support the weight of its enormously heavy belly region. The back of* Deinonychus *is similarly arched, but its strength is necessary to support the powerful shoulders.*

Tails

The tail of Deinonychus *serves as a dynamic stabilizer and counterbalance, while that of* Brachiosaurus *anchors the leg muscles and the tendons of the neck, and acts as a radiator to prevent overheating.*

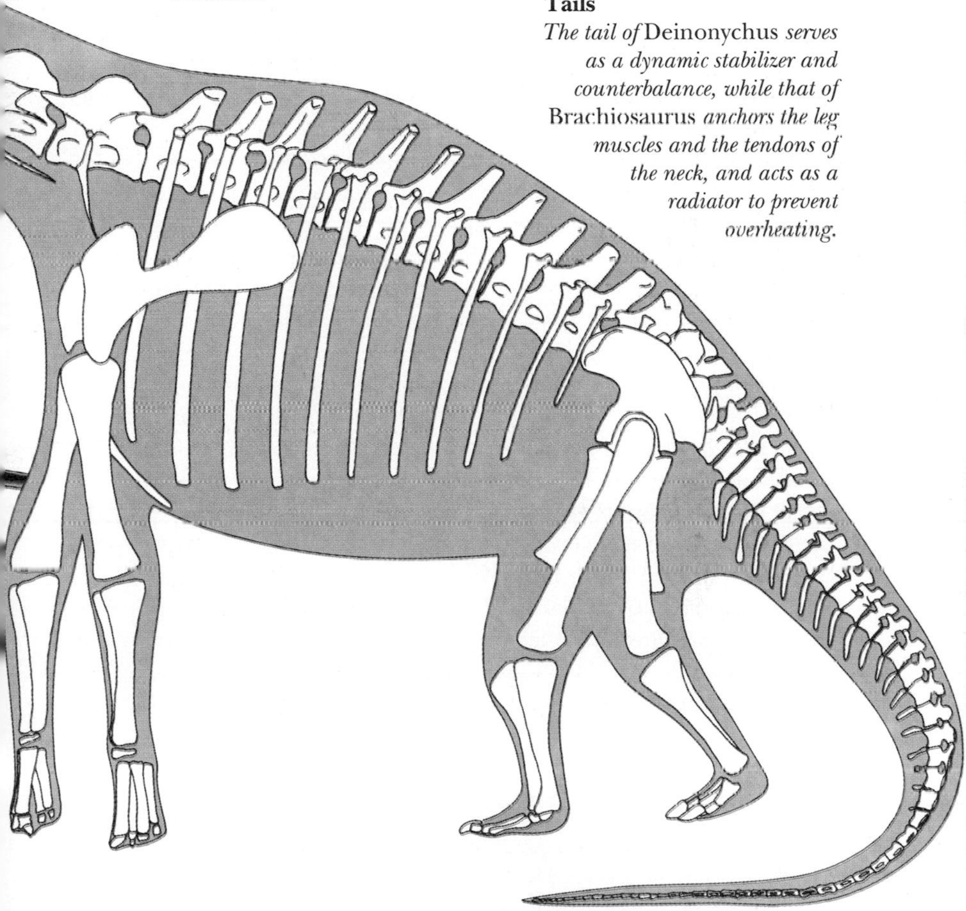

est activities such as walking for any length of time would have generated large quantities of surplus muscular heat, causing the body temperature to rise quite sharply. Dumping this heat sufficiently quickly may have presented severe problems. The same problem is encountered in elephants today, and these use their very large ears as radiators. Brachiosaurs may have used other techniques for which we have some limited evidence. The air sacs could well have been used as heat dumps. The long cylindrical neck and tail would also have acted as quite effective radiating surfaces, especially if the blood could be selectively channeled under the skin, as it can in modern reptiles. Another feature, not so far mentioned, which may also have had a role in temperature control is found in the head. The nostrils of these dinosaurs are exceptionally large; this is purely speculative, but it is at least possible that one of the reasons for the huge nostrils was to allow for blood to be cooled by air passing across the delicate membranes inside them. This certainly provides a more logical explanation for the size of the nostrils than any other that I can think of at the moment.

Deinonychus, with its small size and rather spindly construction, could not have kept its body temperature stable in the same way as a brachiosaur. Yet looking at the general biological attributes of such a theropod makes it hard to believe that such an animal could operate efficiently with anything less than stable internal body temperatures. The brain of this dinosaur was large by reptile standards – close to that of some modern birds. The eyes were large, indicating acute vision; the ear was well developed; its sense of balance must also have been extremely acute, given the style of life, as a running biped, and the offensive tactics which this animal used in order to capture and kill its prey. Add to this the highly cursorial construction of the skeleton and the probability that it had a high-pressure circulatory system, and what emerges is a sophisticated, finely balanced, fast-moving, intelligent predator.

All these attributes require a high level of central control over the activities of the creature which imply a stable internal body temperature. Only in such conditions could the fine control systems necessary to operate such an animal have developed. The implication must be that they had developed an alternative temperature control system depending on a fast metabolism: some measure of endothermy, like that of mammals and birds.

MUSCLES OF *DEINONYCHUS*
The small and powerful skeleton of Deinonychus *provides a good opportunity to show the process of muscle reconstruction in a dinosaur. Starting with the bones, which can be seen in the tail, first the ligaments and tendons can be added, then the muscles and finally the skin is added to give a full, life-like image.*

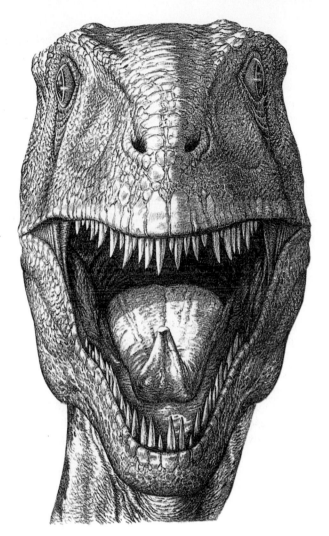

The prey's eye view of Deinonychus *reveals the formidable array of teeth that lines the jaws and the overlapping fields of vision of the eyes, enabling them to focus very clearly on their prey. With these specializations and a devastating turn of speed, these are among the most sophisticated of all predatory dinosaurs.*

Dinosaurian economics

Brachiosaurus and *Deinonychus* seem to be pursuing radically different policies for survival within the design constraints of being a dinosaur.

Large size brings several benefits. Bulk provides safety – it seems extremely unlikely that any predator would have been able to take on an adult brachiosaur. Large size brings with it the benefit of a relatively constant internal temperature, without the need for fast metabolism associated with endothermy and the huge energy cost of feeding an endothermic system. Large size brings economy in metabolism because the energetic requirements of large animals are, in terms of body weight, less than for small animals; large herbivores can survive on much poorer-quality food than small ones. The large ratio of volume to surface area also reduces the risk of dehydration in dry conditions.

There are also costs associated with being very big. Most obviously an enormous amount of growth has to occur before an animal eventually becomes full-sized and invulnerable. Movement is also extremely costly metabolically simply because of the effort to move 20 tons or more of flesh. Agility is also lost. Growth must have put a huge strain both on the metabolism of the animal and on local resources for any animal that must have started life weighing approximately 45 lb (20 kg), and grew to be an adult weighing 20 or 30 tons. No matter how quick growth was, the young would have passed through an extended period of extreme vulnerability to predators. At present no one knows whether brachiosaurs nurtured their young to help them survive to adulthood or whether they were left to fend for themselves.

Small size brings with it a number of survival factors, but relatively few economies. Being small can allow an animal to be fast and more responsive to local changes. The mechanics of building a small, as opposed to a large animal, are less demanding since there is less risk of mechanical failure. Growth is also much quicker, since there is usually much less difference in size between the young and adults in a small species than in a large one. Parents also do not need to invest so much effort in looking after their young. Rates of reproduction can be faster with small species, and the habitat can support a larger number.

Small size also has its costs. There is a greater risk from predators. Small size leaves the body more open to changes in the external environment; water loss and temperature change can be far more stressful to small animals than to large ones. Smaller animals tend to live on tighter energy budgets than large ones, needing more food in relation to their size. Smaller species therefore tend to be much more selective feeders than large ones, which can make them vulnerable to local environmental change which might have little effect on larger species.

Was the comparison useful?

There do seem to be several features which make dinosaurs special. Both *Brachiosaurus* and *Deinonychus* have bodies which, from a mechanical viewpoint, are extremely well supported and very finely tuned to their own way of life. In addition their soft anatomy seems to have had many features in common, most important among these being a high-pressure circulatory system, an efficient respiratory system, and well developed senses and control mechanisms. Both were clearly active creatures.

However, the crucial control of body temperature may have been solved in different ways. The brachiosaur's huge size made it quite easy to stay warm. The deinonychosaur was obliged to use other methods. Since it seems to have developed complex and sophisticated control systems, evidently it must have had some form of fast metabolism which would have provided a constant supply of heat from within the body in addition to the heat generated by muscular activity. This endothermic system is a much more costly one to run, and these costs must be offset by benefits such as increased efficiency of hunting.

UNDERSTANDING DINOSAURS

The story of the dinosaurs seems to be one of spectacular success in their dominance of the Mesozoic world, followed by spectacular failure in their extinction at the close of the Cretaceous Period, leaving only their lineal descendants the birds. We have studied their history and how they were discovered. What conclusions can we draw from this, and how can we understand the group as a whole? I can think of no better way of doing this than by a quick survey of the dinosaurs' time on Earth, with their beginnings in the Late Triassic, following their rise to dominance, and ending with their final demise 66 million years ago.

Origins

As far as we can tell, dinosaurs arose at some time during the Late Triassic. Unfortunately our chances of finding the first dinosaur, or even of recognizing it as such if we did, are extremely slim. Why they arose at this particular time has been and is still a matter of intense debate.

Small dinosaurs such as Coelophysis, *because of their large skin area, are most susceptible to temperature changes.*

The earliest skeletal remains of dinosaurs known are of small- to medium-sized, highly agile, bipedal carnivores. They are recognized primarily through the structure of their legs and hips, which show the mechanical arrangements for a fully upright posture. Trackways of animals walking in the way that dinosaurs would be expected to walk are known from the Middle Triassic, so it is possible that they were around in even earlier times. At the time of their first appearance dinosaurs were, from a purely mechanical point of view, quite advanced reptiles compared with the majority of others living during Triassic times. The way that their limbs worked allowed them to move quickly, or conversely to carry great body weight efficiently. This potential was rapidly translated into dominance by dinosaurs of all land communities. The precise way in which they achieved this dominance has been much discussed.

My preference, among the many current theories, is one which allows the dinosaurs to prosper simply because they were reptiles, and is linked to the environmental conditions which may have prevailed in the Late Triassic world. It is probable that climatic conditions in Late Triassic times were considerably warmer than at present. There were certainly no ice-capped polar regions, and many areas of the world appear to have been arid. If these were indeed the prevalent conditions, then they may well have favoured reptiles on land. Reptiles were one of the earliest groups of land dwellers to solve two of the three main problems involved with life on land: supporting the body against the effects of gravity; preventing water loss in air; and buffering the body against rapid changes in the temperature of the air. This is clearly an artificial set of problems because they are inspired by our perception or preconceptions. Nevertheless examining them may help to clarify some of the issues.

To solve the first problem, reptiles developed well designed skeletons for support on land. The second came through the development of a thick, scaly skin which cut down water loss through the skin, and of methods for reducing to a minimum water losses from the body in urine by excreting a paste composed of salts of uric acid. The third problem was never satisfactorily solved by reptiles, but was eventually solved by mammals and birds through the development of endothermy, combined with insulation to control the rate of heat loss.

Given this set of attributes, the Late Triassic may well have been a time of great potential for reptiles generally. If the world was predominantly warm, dry, and sunny, conditions would have been well suited to animals which relied on external warmth to maintain their body temperatures, were able to minimize water loss in arid conditions, and were able to live on a rather poor or patchy food supply. Endothermic mammals which appeared at a similar time in the fossil record share a set of attributes which may not have suited them as well to the conditions of the time. Endothermy, though clearly advantageous in some circumstances, brings with it considerable costs: it requires a large intake of food as fuel to generate internal body heat, and of large volumes of water which are released by panting or sweating to keep cool, so that it is least effective in hot conditions. Almost all we know about Late Triassic environments seems to put mammals at a disadvantage and favor the standard ectothermic reptile.

From these considerations we may reasonably suppose that Late Triassic and later Mesozoic conditions were better suited to reptiles than to mammals. Whether dinosaurs in particular managed to dominate all others because they were the lucky survivors of some sort of environmental catastrophe remains to be proven convincingly, but the fact that they maintained their dominance in later times is probably due to the prevailing climate. It was also undoubtedly linked to the fact that among reptiles, dinosaurs combined mechanical perfection for life on land with some key physiological innovations which allowed them to be extremely active; most notable among these were the high-pressure blood circulation made possible by the fully divided heart, and highly efficient lungs. These features in particular allowed dinosaurs not only to be active, but also to sustain their levels of activity far longer than modern reptiles are able to.

Dominance

By the close of the Triassic dinosaurs were so numerous that their remains are found in considerable abundance in several parts of the world, indicating their rapid rise to dominate life on land. With the rise in abundance, their variety also increased greatly to include small and large carnivores as well as an enormous variety of small, and medium to very large, herbivores showing a huge range of adaptations to their diet of plants. Beyond the Late Triassic very warm, mild, practically seasonless conditions seem to have prevailed, coupled with an increase in the availability of water; this seems to have brought a great enrichment of the plant life and of the dinosaurs to feed upon the vegetation.

Not only did dinosaurs become anatomically varied, but they appear to have developed a variety of sophisticated behavior patterns seldom seen in modern reptiles. Some became very accomplished predators capable of pursuing and despatching their prey with as much ease as the great cats of today. Others lived in herds and developed cooperative social behavior which seems to show strong parallels with that of modern herding mammals. Evidence from nesting sites suggests not only colonial breeding, but complex parental care of the young in some species, and the likelihood of transcontinental mass migration equivalent to that of modern wildebeest and reindeer.

Dinosaurs' apparent high activity and sophisticated behavior have been interpreted as strong evidence that dinosaurs were endothermic like modern mammals and birds. However, the evidence for this is far from convincing for the majority of dinosaurs. The prevailing warm and non-seasonal climate, combined with the unique aspects of dinosaur physiology, allowed them for the most part to combine the economies of an ectothermic metabolism with a warm, constant internal temperature which allowed complex control systems to develop, and circulatory and respiratory systems allowing brisk physical activity.

The only groups which do not fit well with this model for dinosaur success are the small, highly active dinosaurs exemplified by *Deinonychus*. Dynamic small dinosaurs may well have been to some extent endotherms, supplementing their body temperature by internal heat generation in a manner similar to that of modern mammals and birds – though precisely how similar this version of endothermy may have been to that seen in mammals and birds today we will probably never be able to say. This possibility of some smaller dinosaurs being endotherms is given further support by the knowledge that endothermic birds most probably originated from small, carnivorous dinosaurs quite closely related to *Deinonychus*.

It thus seems that during the Mesozoic the dinosaurs generated not only a great diversity of anatomical types, but an equal diversity of physiological types.

Extinction

The fact that dinosaurs became extinct might suggest they had some "Achilles' heel" which led to their eventual downfall. Most realistic scenarios for the extinction of the dinosaurs involve some form of environmental disruption occurring 66 million years ago. The principal questions are: How quickly did the catastrophe come about? Were the dinosaurs among a range of random victims, or selectively killed? My suspicion is that the majority of dinosaurs were killed selectively by environmental deterioration. Global temperatures are thought to have declined quite markedly at the close of the Cretaceous Period, the seasons seem to have become more distinct.

The most obvious drawback of dinosaur biology, according to the models discussed above, is that they were ultimately reliant upon universally mild climatic conditions. The survey work of paleontologists such as Leigh Van Valen and Robert Sloan seems to show that there was a gradual decline in the abundance and variety of dinosaurs during the last 7 to 10 million years of the Cretaceous Period, and that this coincides with a slow but gradual rise in the number and variety of mammals. This would seem to tie in reasonably well with a model for dinosaur extinction prompted by progressive climatic deterioration. That could be perfectly true even if there were a meteorite strike or huge volcanic eruptions near the end of this time.

The huge size range of dinosaurs, from 120-foot-long giants weighing tens of tons, such as "Seismosaurus", through to miniatures weighing a few pounds, make attempts to fit dinosaurs into a particular stereotype virtually impossible.

The big ectotherm worked well in the Mesozoic provided the weather was warm and pleasant for most of the time. Average temperatures in ancient times can be discovered by analyzing sea floor sediments for the proportions of different isotopes of oxygen, which give a record of the amount of biological activity when the sediment was laid down. These studies suggest that average yearly temperatures in the early part of the Late Cretaceous were in the region of 64 to 68F (18–20C). Periods of prolonged cool or cold conditions would have caused a fatal drain on body temperature from which large ectotherms would have had little chance of recovery. Having naked skins, as all these reptiles seem to have had, judged from the skin impressions which are known, means that body heat can be lost quite rapidly through wind chill. Once they were chilled, even basking in the sun would be of little value to the larger dinosaurs, since they were too big to warm up quickly. Indeed the larger a reptile is, the more difficult it becomes to use basking as a means of temperature control. One fact which is not often mentioned in the work of Colbert, Cowles and Bogert on body temperature control in small and large alligators (page 113) is that several of the larger alligators involved in their experiment apparently died of the effects of sunburn after prolonged exposure to the sun. Their skin actual began to bake rather than being able to pass the heat which was being absorbed directly into the blood and around the body.

The small endothermic dinosaur has similar problems to the large ectotherm, again because it lacked a layer of insulating fur or feathers. Even though these dinosaurs were capable of generating body heat internally they would have been unable to control the rate at which they lost heat through the skin to the air. Thus while they could cope with short-term local changes in temperature, global climate deterioration would have affected them just as badly.

If, as has been argued, birds did evolve from small carnivorous dinosaurs with some sort of endothermic metabolism, their survival across the period of extinction 66 million years ago may be linked to the evolution of feathers; these not only enabled birds to fly, but insulated their bodies against heat loss.

Looking at the extinction from the point of view of the survivors, the principal beneficiaries seem to have been animals that were endothermic and well insulated – birds and mammals. Other survivors included a variety of small amphibians, small lizard-like reptiles, and tortoises, which are readily able to survive periods of cold by hibernating in crevices or burrows (impossible for a 20 ton dinosaur!) or live in an environment of relatively constant temperature. For example the freshwater crocodiles may have survived simply because they lived in and around water, which has a high specific heat capacity and would have buffered them against temperature changes in the air.

Dinosaurs were therefore successful at first because they were reptiles, and could live economically in warm, dry climatic conditions, and yet at the same time they had developed an efficient heart and lungs which allowed them to be highly active. Their success was ensured through the Mesozoic because this period favored animals that were adapted to a generally warm, non-seasonal climate. These conditions permitted the evolution not only of the huge sauropods which kept warm by sheer size, but also small dinosaurs which had a supplementary mechanism – though this was not effective enough to cope with prolonged cooling. The close of the Mesozoic Era ushered in cooler, more variable climatic conditions unsuitable for animals designed to operate at a constant body temperature but lacking any effective insulation. These more variable conditions well suited the mammals and birds which combined endothermy with insulation and evolved rapidly after the end of the Mesozoic.

In short, dinosaurs were simply the best solution that Nature could come up with for the particular combination of conditions which existed on Earth during most of the Mesozoic. Dinosaurs were right for their time but, as always, times change.

ABOVE: *As part of the series* Dinosaur! *a series of life like models were created for animation sequences by studios in Bristol and Manchester (England). Their life-like appearance, as this model of* Centrosaurus *shows, is a tribute to the skills of the model makers. Turning these models into "living" animals was the work of Peter Phillips.*

RIGHT: *Given the right evolutionary circumstances, dinosaurs might have evolved into sophisticated technological creatures.*

CHAPTER NINE

The Making of Dinosaur!

Showing dinosaurs on television has a severe problem – real dinosaurs don't move. We can describe them in loving detail, we can draw their bones, we can even paint pictures or make models of them, but we can never see one alive. They are ideals subjects for books, but television is a dynamic medium suited to natural history presentations where animals can be seen to "perform," while the viewer, and for that matter the commentator, can sit back and be enthralled by entirely natural stars. The problem with dinosaurs is how to make them "come alive" both as animals and as a subject. They have their own fascination, as shown by the success of many lavishly illustrated books on them, but that in itself can lead to impossibly high expectations on the part of the viewer. Presented on television, they may disappoint rather than inspire.

To avoid these pitfalls the production team for the *Dinosaur!* series took a radical approach to the subject – one that is not reflected in the structure of this book. Each of the four episodes has a broad, but distinct, theme: "Tooth," "Bone," "Egg," and "Feather."

"Tooth" is concerned primarily with the wonder and excitement created by the discovery of dinosaurs. It all started with the discovery of the teeth and bones of *Megalosaurus* in 1817, and continues to this day, for example with the finding of the new *Tyrannosaurus* skeleton in Montana. The early history of dinosaur research was dominated by the discovery of teeth, from the first ones discovered in Britain to

The American sculptor Jim Gary has made some extraordinary metal "dinosaurs" out of the scrapped remains of automobiles! This is his exuberant tyrannosaur.

those discovered in the American Midwest, and culminating in the marvelous dinosaur discoveries of the 1870s in Europe and America.

"Bone" takes the story from the discovery of the first skeletons of dinosaurs, particularly those discovered during the American "bone wars" toward the close of the last century, and shows how the marvelous discoveries that were made have allowed paleontologists to gradually piece together many of the mysteries which surrounded this group of animals.

"Egg" presents dinosaurs as complex living creatures. Eggs have provided us with a profound insight into dinosaurs' social and family life. The first, discovered in the 1920s, not only proved that dinosaurs laid eggs, but revealed something about their growth and sexual differences. However, it was not until the mid-1970s, when new nest sites and embryo bones were discovered in Montana by Bob Makcla and Jack Horner, that a whole new insight into the unexpected sophistication of dinosaurs' way of life began to emerge.

"Feather" is really two subjects knitted together, for a definite reason. Dinosaurs and birds seem to be very closely related, judged by research carried out over the last fifteen years – birds are, in other words, the nearest living relatives of dinosaurs. It was the preservation of feather impressions around fossil skeletons which led ultimately to this view. The closeness of the link is examined, which leads naturally to the question to which everyone wants an answer – why did the dinosaurs become extinct? And, if birds are close relatives of dinosaurs, why did they survive while dinosaurs did not? These questions lead to a consideration of what made dinosaurs tick, to what extent they were like modern reptiles or birds, and the contributing factors that cause extinctions of large groups of organisms at intervals throughout Earth's history.

Within each of these episodes are interwoven a number of plots and subplots, mostly involving scientific detective work and historical themes. In order to avoid simply delivering a series of four long, albeit colourful, lectures, a wide variety of presentational techniques have been used; these range from straightforward commentary or interviews with experts and enthusiasts, through participation in dinosaur excavations on location, to historical dramatizations and accurate animated dinosaur models which simulate the actions of living dinosaurs.

INTERVIEWS

The series would not have been possible without the willing cooperation of paleontologists, most of whom are colleagues and friends, from various countries. The fact that scientists give their time freely to interviewers, despite the enormous disruption that a television crew can bring, is not often appreciated by the television audience. Few of them are natural television interviewees, most preferring to carry on with far more important, and time-consuming, scientific research. Yet they all willingly helped and contributed whenever possible, and brought with them their dedication and their scientific values and training, mixed with a quite genuine sense of excitement and even of fun – essential ingredients in the work we do. And that does not apply just to young, enthusiastic researchers. Eminent professors such as John Ostrom and Stephen Jay Gould convey as much wonder and thrill as any. To them all I would like to say, as publicly as possible: Thank you.

The excitement of being involved in dinosaur research is admirably conveyed by the evident enthusiasm – almost to the point of babbling incoherence – of some research workers as they talk with the series presenter, Walter Cronkite, about what fires their enthusiasm for the subject. In similar vein Jim Gary, an extraordinary character, waxed lyrical about making dinosaurs from the broken parts of wrecked cars. We also get a feel for the excitement of a brand-new excavation of a skeleton of the most fearsome of all dinosaurs, *Tyrannosaurus*, as it is slowly and painstakingly coaxed from its rocky tomb in the badlands of Montana; and begin to wonder at the meaning of the extraordinary dinosaur tracks left by a dinosaur such as *Coelophysis*, which form a translantic link between America and Britain. But while it is all very well to enthuse about dinosaurs nowadays, how did we get to know about them in the first place? This brings us to the theme of the story – a tooth.

HISTORICAL DRAMATIZATIONS

The history of dinosaur studies truly began with the tooth that Canon William Buckland, cleric and geologist of Christ Church College, Oxford, pondered over when he was first handed some large fossil bones in 1817. The tooth, with a few other bones, collected by some quarrymen belonged to the first dinosaur ever to be described, *Megalosaurus*. The tooth was a source of wonder and mystery to Buckland. It clearly belonged to some ancient carnivorous creature – but of what kind? With the help of Baron Georges Cuvier, Buckland was able to deduce that *Megalosaurus* was a giant predatory reptile of a bygone age. Here was something utterly unlike animals living today and, to Buckland's clerical mind at least, one of the many creatures which failed to find refuge in Noah's ark and perished in the Flood.

Filming at Brasenose College, Oxford, with our own "Dr Buckland," we showed him examining an actual specimen of *Megalosaurus* studied by the real Buckland and Cuvier. We tried to

This may not be the actual Baron Georges Cuvier, but having actors to play such roles can greatly enliven what might otherwise be a relatively uninteresting historical account.

convey some of the contradictions of his being a man of the cloth, and at the same time a geologist studying the history of the Earth.

Another dramatization of Gideon Mantell and his discovery of *Iguanodon* is also included in this episode. In fact the story of Dr Mantell's wife finding the first teeth while her husband was making a house call is probably a romantic fiction. But the value of Mantell's pioneering work should not be underestimated. He was the first to attempt the reconstruction of a dinosaur skeleton with his sketch of *Iguanodon*.

ANIMATION

The drama of these early discoveries is given added life by the use of animation sequences. *Iguanodon*, one of the first dinosaurs to be recognized, is brought to "life" and seen by the viewer browsing on trees in an Early Cretaceous scene and drinking at a water hole. The realism of the scenes is quite stunning and the tricks that made it all possible are explained below.

Models for filming

The majority of the models used in the various film sequences were made in England by Peter Minister in his studio in Manchester, with some additional models (the *Centrosaurus* on page 266 and the *Diplodocus* head on page 235) made by Lyons Model Effects Ltd in Bristol. Both start with accurate scale drawings, and then the dinosaurs were sculpted to reduced scale (between 3 and 5 feet in length) in "Plastilina" wax on a simple metal armature. These were checked for accuracy by the scientific adviser to the series. The completed model was cut up into sections – head, neck, torso, tail, arms, and legs – so that it could be molded, using a very high-quality plaster mixture, "Crystacast."

Once the model was cast, the wax was removed and used to make other dinosaur models for the series. The plaster molds were the basis of the models to be used for animation. First the molds were left for two days to dry and harden completely. Then a specially designed inner core was made to house the robotics that would be fitted in the final model.

The hardened molds were filled with foamed latex, and the inner core carefully maneuvered into place. Next they were baked in an oven for 3 hours at 212°F (100°C), setting the foam to an extremely light, spongy, and skin-like texture. Each of the molded sections was then glued together around the robotic mechanisms, which included a radio receiver, servo-motors, and piping to operate inflatable bags used to simulate breathing and other movements. The seams on the

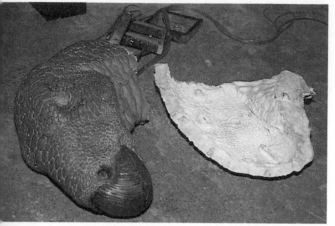

BRINGING DINOSAURS TO LIFE.

LEFT: *"Crystacast" molds were used to cast the foamed latex head of* Iguanodon *which was then painted and the mechanical devices inserted.*

RIGHT: *In the film studio the head is made to perform, in this case by puppeteers using control rods, cables and air tubes. This provided the real-time movements for some scenes.*

Coordinating movements between different parts of the body required a great deal of practice before what was seen through the camera viewfinder began to look "natural". In this case the head of Iguanodon was being operated to simulate chewing plants, while the hand was being used to show how it could grasp vegetation and push it into the mouth using the mobile fifth finger.

Iguanodon *is seen here as it appeared in the series, apparently standing in shallow water at a watering hole. What this involved was the filming of a scene outdoors in which a man stood in a pond shuffling his feet to create ripples. Film studio shots of the robotic* Iguanodon *model were then edited into the scene and the man's image erased.*

RIGHT: *The robotic model of* Iguanodon *as it appeared in the studio before being placed in the pond scene (above).*

molded pieces were trimmed and filled until no trace of a join was visible. Each dinosaur was then painted, using specially flexible paint, and the finishing touches added, including the teeth, claws, and eyes.

Robotics allow a model animal to move in a wide variety of very realistic ways. They are operated from remote control consoles, or with mechanical pushrods mounted on small hand-held frames, or by blowing in tubes which operate bellows within the body of the animal. All this had to be intensively rehearsed in preparation for the filming sessions.

Filming of the models was directed by Peter Phillips, and took place in a model studio, usually against a flat blue background to allow the filmed sequence of the model to be superimposed imperceptibly onto a background film of real scenery by the "Chroma-Key" process. One of the most important features of such filming is the lighting, which is set by a cameraman specializing in model and special-effects photography. The lighting used on the model has to match that of the background scene. Usually "gobos," which are masks cut out in various shapes, are used to create realistic shadows across the subject. A gobo in the shape of a tree can give the impression that a tree in the real scene is actually shading the dinosaur model. It is vital to give the effect of the model being a real part of the scene, rather than something that has been very obviously stuck on top – a feature which the human eye has an extraordinary knack of spotting unless great care is taken.

In addition to the lighting, the camera angle has to be set exactly right so that the model and real scene are viewed at the same angle. Videotape is used to view the scenery and the model in preview before actual filming begins. This ensures that the lighting, depth of shadows, and size are correct, and that the model sits in the background scene comfortably.

The puppeteers, of whom there were usually between three and five, rehearsed each scene with the director and cameraman until everyone knew exactly what was expected of them. Coordinating five people to create the behavior of a single creature was sometimes extraordinarily difficult. When everyone was ready the filming began, usually involving three or four "takes" before everyone was completely satisfied with their performance. Video recording again made it possible to view the film immediately after the "take," instead of having to wait until the following day for the film to be processed. This greatly improved the speed and efficiency of the entire process and allowed the team to move confidently onto the next scene without any fears of having to do the same scene again.

Rather than using the "stop-frame" animation popular many years ago – in which animation is created by filming literally frame by frame, with the dinosaur model being moved by a tiny amount with each frame as a cartoonist does in animating cartoon characters – we used "real time" animation. This means that the models are moved at real speed, by their controllers, giving smoother movements, and a slight blurring of the image caused by the movement of the model as it is filmed. The viewer's eye is used to such blurring in scenes showing real objects, so the effect helps to make the movement look realistic.

The whole process sounds, and seems when it is being done, incredibly tedious and artificial, but it is surprising how realistic these animations can appear once they have been edited and supported by a soundtrack with appropriate noises, commentary, and music.

Large animated models

The process by which Peter Minister created his miniature robotic dinosaurs is very similar in most respects to that used in the large animated models seen today in museums and parks. However, the extra space inside a full-sized dinosaur model allows for more and better robotic equipment.

The full-sized *Allosaurus* "Alice" constructed by Robbie Braun and also featured in this series is in most respects a larger, more durable, and more sophisticated version of one of Peter Minister's models. Operated remotely by museum visitors from a push-button panel, the robotics inside this dinosaur model are driven by compressed air through a computer-controlled system of valves. This system clearly impressed Ray Harryhausen, one of the most gifted film animators of dinosaurs, who produced the magnificently realistic movements of dinosaurs in the movie *Valley of the Gwangi* using only a small, flexible model, which he manipulated into "life" by the stop-frame technique explained above.

Skeleton animation

Elsewhere in the series another and quite different technique was used to bring life back to dinosaurs. In this instance our work centered on the wonderful skeleton of *Brachiosaurus* in the Berlin Museum of Natural History. As I have emphasized in this book, it can be instructive to look at the extremes of dinosaur design in order to understand how dinosaurs in general may have worked as living creatures – and *Brachiosaurus* is in every sense extreme at 75 ft (23 m) long, over 34 ft (11 m) high, and weighing perhaps 20 or 30 tons in life. It simply dwarfs *Diplodocus* – the classic gigantic dinosaur of so many museum displays – which stands alongside it in the museum, and reduces even an experienced dinosaur paleontologist such as myself to hushed tones of awe and reverence.

We worked from my own descriptions of the parts of the soft anatomy of the dinosaur which were of interest, and then literally painted the anatomy into the skeleton. This allowed us to combine the wonderful artistic skills of John Sibbick, who painted the soft tissues on sheets of paper, with the film editing and animation skills of Peter Phillips.

As the huge muscular heart, the great blood vessels, and the lungs are described, so they appear within the skeleton. In another passage, feeding and the structure of the stomach are discussed in some detail, and to support this the stomach and intestines appear as if by magic, making the point far more clearly and with greater appeal to the imagination than a complicated description in words could ever do.

Finally as the *pièce de résistance*, the muscles of a dinosaur are reconstructed, again using *Brachiosaurus*. First the areas on bones that show

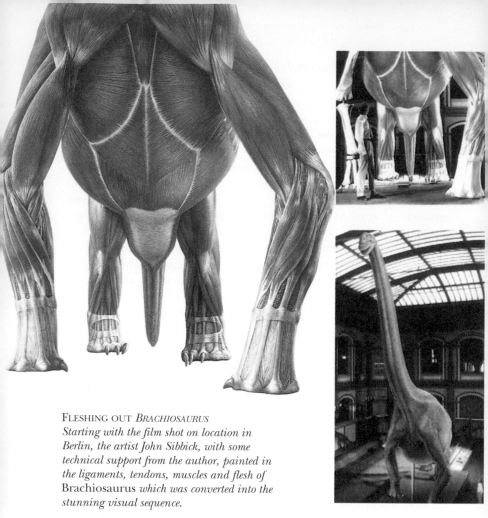

FLESHING OUT *BRACHIOSAURUS*
Starting with the film shot on location in
Berlin, the artist John Sibbick, with some
technical support from the author, painted in
the ligaments, tendons, muscles and flesh of
Brachiosaurus *which was converted into the*
stunning visual sequence.

signs of muscle or ligament attachment are described and demonstrated on the skeleton. Then, using John Sibbick's artwork, Peter Phillips' editing skills, and some scientific advice the ligaments, tendons, and muscles are added. Finally the skin is added, and the animal comes to life and leaves the museum – a symbolic reconstruction of an animal from the past, which is in effect the culmination of the scientific work of any paleontologist.

I have tried with the aid of John Sibbick's marvelous skills to create the same symbolism of the work of dinosaur paleontologists in the illustration which adorns the cover of this book. The choice is very deliberately *Deinonychus*, whose discovery was a key event in our reawakening to the enigma of the dinosaur. It starts as a paleontologist would, with the bare bones and culminates in the representation of the living dinosaur.

Epilogue

Dinosaur research has not stood still in the two years since the original version of this book was written. Indeed, interest in dinosaurs has been maintained for a variety of reasons. Unfortunately, to my mind, some of the interest surrounding dinosaurs has not been entirely beneficial. One of the most notable headline-catching incidents has been the controversy surrounding the discovery and excavation of a new skeleton of the dinosaur Tyrannosaurus, nicknamed "Sue." This unfortunate individual has become the center of a bitter legal dispute about ownership, which has exposed some of the raw nerves associated with the excitement of such discoveries.

Almost since the first fossils were unearthed they have been valued, either as simple mementoes of a visit somewhere special, or for their intrinsic beauty, or in a financial sense, as possessions that may have monetary value. Sadly, the commercialization of fossils, and dinosaurs in particular, in recent years has made them victims of their own success – or perhaps Professor Richard Owen's success. It was, after all, Richard Owen who dreamed them up all those years ago. I cannot help but be filled with sadness when I read in the papers that a new dinosaur skeleton, such as the one mentioned above, is hitting the headlines because of a dispute which has at its core commercial interest: "How much is it worth? What could we get for it?" I may be naive, but I believe that fossils should not really belong to anybody to buy or sell. Surely they are part of the history of our world; they are actually a tangible connection with the past. In almost every case the minerals in the fossilized skeletons were laid down by the animal when it was alive. In truth, fossils are a part of our natural heritage – or at the very least of the nation in which they are found. Ultimately fossils should find their way into museums where everyone should be able to enjoy them.

Another interesting controversy concerns Professor Owen and the naming of the dinosaurs. Dr Hugh Torrens of the University of Keele, a noted geologist and historian of science, has recently pointed out that I (and a few others!) have been in error when claiming that the dinosaurs were "born" in 1841 at the famed meeting of the British

Association for the Advancement of Science in Plymouth. It appears that none of the reports of the meeting give any mention whatsoever of Owen having uttered the name "dinosaur" during his second discourse on the fossil reptiles of Britain. In fact, it seems likely that he did not think of categorizing some of the giant fossil reptiles as "dinosaurs" until some time early in 1842, while he was writing up his own report of the meeting.

The last year or so has seen some interesting reports of new discoveries. Dr Paul Sereno from the University of Chicago has been involved in the discovery of a crop of new, very early dinosaur skeletons in Argentina: *Herrerasaurus*, a previously known, but poorly described carnivorous dinosaur; and *Eoraptor* ("dawn thief"), another extremely early dinosaur, which is completely new to science. The latter, a very small, highly agile, predatory creature, seems in many ways to be very close to what we expect the very earliest dinosaur to look like. On the Isle of Wight in southern England, a very well preserved skeleton of a brachiosaur-like dinosaur has been found by Steve Hutt of the Sandown Museum. This is the first time that any reasonable remains of a dinosaur like this have been found in Britain, so it is both an exciting and important discovery. Indeed the Isle of Wight is becoming once again a treasure trove for dinosaurs. A beautifully preserved carnivorous dinosaur was dug up there quite recently and, less than a month ago, a talented amateur collector named David Cooper picked up a rolled "rock" from the seashore which he presented to me. I was startled to discover that it was the first reasonable portion of the skull of an ankylosaur (armored dinosaur) from Britain.

Dinosaurs may be things of the past, but they are stimulating a great deal of interest. An organization called "The Dinosaur Society" has been founded as a non-profit-making organization seeking to support dinosaur research workers in various ways throughout the world. Research groups are currently undertaking important projects in order to describe new discoveries, to understand better the evolution of the group, and their relationship to other animals and plants of the time. There may not be that many dinosaur research workers working worldwide, but we are trying hard to clear away some of the mists that shroud our understanding of the past. However, we very much appreciate the fact that so many of you enjoy and can share, through books such as this, our interest in those terrible old lizards.

DAVID NORMAN
Cambridge, England, February 1993

FURTHER READING

The history of paleontology

Edwin H. Colbert, *Men and Dinosaurs: The search in field and laboratory*
(Evans Bros, London, 1968). One of the classic books on the history of
dinosaur discoveries and the personalities of the people involved,
written by an eminent man whose own life and experiences just overlap
those of some of the early workers of the twentieth century. Perhaps a
little dated now, and quite difficult to find outside public libraries.

Adrian J. Desmond, *The Hot-Blooded Dinosaurs: A revolution in
palaeontology* (Blond & Briggs, London, 1975). Strongly partisan book
pushing the "hot-blooded" theory, but highly enjoyable on the
history of dinosaur discovery and scientific interpretation, especially
that of physiology.

Martin J. S. Rudwick, *The Meaning of Fossils: Episodes in the history of
paleontology* (University of Chicago Press, Chicago, 2nd edition 1985).
Extremely interesting and well written, somewhat more academic
book on the history of paleontology. An important work, but not
specifically on dinosaurs.

John N. Wilford, *The Riddle of the Dinosaur* (Alfred A. Knofp, New
York, 1985). Highly readable account of the history of the study of
dinosaurs, but sadly lacking in illustrations.

For the enthusiast

Alan J. Charig, *A New Look at the Dinosaurs* (Natural History Museum,
London, 1979). Careful and authoritative review, weak on the variety
of dinosaurs and a little outdated now.

David Norman, *The Illustrated Encyclopedia of Dinosaurs* (Salamander,
London, 1985). Detailed and extensively illustrated.

Robert T. Bakker, *The Dinosaur Heresies: New theories unlocking the mystery of dinosaurs and their extinction* (William Morrow, New York, 1986). Summary of Bakker's very particular views about dinosaurs, their world, and the work of other paleontologists. Lively, but too often the science suffers at the expense of entertainment.

Dinosaurs Past and Present, edited by Sylvia J. Czerkas and Everett C. Olson (Natural History Museum of Los Angeles County, Los Angeles, 1987, 2 volumes). Fascinating collection of articles by paleontologists and dinosaur artists, exploring many topics.

Gregory S. Paul, *Predatory Dinosaurs of the World* (Simon and Schuster, New York, 1988). Heavily influenced by the author's own views. He is first and foremost a dinosaur artist, and the book is fascinating for its extraordinary illustrations.

Francine Martin and Pierre Bultynk, *The Iguanodons of Bernissart* (Royal Institute of Natural Sciences, Brussels, 1990). Short booklet describing the discovery of *Iguanodon* in Belgium, and subsequent studies of the animal. Updated English-language version of a longer book: Edgar Casier, *Les Iguanodons de Bernissart* (1960).

The Dinosauria, edited by David B. Weishampel, Peter Dodson, and Halszka Osmolska (University of California Press, Berkeley, 1990). Authoritative review by a wide range of dinosaur paleontologists. Extremely detailed, rather heavy going.

The Dinosaur Data Book, produced by Diagram Visual Information, in collaboration with a panel of paleontologists (Natural History Museum, London). Compendium of facts and statistics.

For the general reader

John R. Horner and James Gorman, *Digging Dinosaurs* (Workman, New York, 1988). Personal and interesting review of Jack Horner's discovery of dinosaur nesting sites and research into their behavior.

David Norman, *The Prehistoric World of the Dinosaurs* (Bison, London, 1989). Well illustrated book on the life and times of dinosaurs.

David Norman, *A Discovery Guide to Dinosaurs* (Salamander, London, 1989). Small-format, somewhat simplified version of *The Illustrated Encyclopedia of Dinosaurs*.

David Norman and Angela Milner, *Dinosaur* (Dorling Kindersley, London, 1989). Simple, clear photographic guide to dinosaur fossils.

Michael Benton, *On the Trail of the Dinosaurs* (Grisewood & Dempsey, London, 1989). Clear and well written introduction to the world of dinosaurs and the work of paleontologists.

Jean-Guy Michard, *Le Monde perdu des Dinosaures* (Gallimard, Paris). In French. Pocket-sized book with a wealth of accurate information.

For younger readers

David Norman, *The Spotter's Guide to Dinosaurs* (Usborne, London, 1980). Rather old now, but pocket-size and easy to understand.

David Lambert, *St Michael's Dinosaur Quizmasters* (Grisewood & Dempsey, London, 1983). Well presented question-and-answer book.

John R. Horner and James Gorman, *Maia: A dinosaur grows up* (Museum of the Rockies, Bozeman, Montana, 1985). A wonderful short book about how the dinosaurs were looked after by their parents.

David Norman, *The Age of the Dinosaurs* (Wayland, Hove, England, 1985). For younger children.

David Norman, *Dinosaurs* (Salamander, London, 1988). Large-format poster book with simple text and illustrations.

Creatures of Long Ago: Dinosaurs (National Geographic Washington, D.C., 1988). Beautiful little pop-up book about dinosaurs. Easily the best of its kind.

ACKNOWLEDGMENTS

Picture and Artwork Credits

Boxtree would like to thank all the artists, museum archives, private individuals and picture agencies who have generously provided the pictures and artwork in this book, and by courtesy of whom they are reproduced. Credits are by page number.

American Museum of Natural History, Department of Library Services: 105 bottom (Trans. no. 2626) and top (Trans. no. 3141), 110 right (Neg. no. 125158); **Ann Ronan Picture Library:** 12, 75; **Donald Baird:** 16 top left and right, 17, 18 (Princeton University); **Robert T. Bakker:** 91, 132–33; **W.H. Ballou** (from *The Century Illustrated Monthly Magazine,* New York, 1897): 71, 98; **Bavarian State Collection:** 78; **Bayerische Staatssammlung für Paläeontologie und historische Geologie,** Munich: 194 left; **Denise Blagden:** 150, 151, 152–53, 254 right, 254–55; **Board of Commissioners of the Central Park** (from the *Twelfth Annual Report,* New York, 1868): 76; **Brigham Young University:** 19 (Mark A. Philbrick); **British Architectural Library RIBA,** London: 14–15; **British Geological Survey:** 31; **A. Buckham:** 217; **Bürgermeister Müller Museum,** Solnhofen: 211; **Michael Coates:** 48, 51, 147, 149, 198, 202; **Simon Conway Morris,** University of Cambridge: 89; **Penelope Cream:** 122 (after John H. Ostrom), 200 (after Dr Sankar Chatterjee); **M.L. Dollo:** 194 bottom left (from *Bulletin du Musée Royal d'Histoire Naturelle de Belgique,* III, Brussels, 1884), 239 (from *Bulletin Scientifique de la France et de la Belgique,* XL); **Angelika Elsebach:** 62–3; **Vivien Fifield Picture Library:** 152–53; **B. Faujas-Saint-Fond** (from *Histoire Naturelle de la Montagne de Saint-Pierre de Maastricht,* Paris, 1799): 68; **Field Museum of Natural History,** Chicago: 102–3 (CK 9T); **Brian Franczak:** 106, 158, 171, 188–89, 259, 262–63; **Ruth M. Goldwyn:** 11 (artist unknown), 160, 216, 269, 270; **Granada Television:** 273 top, 276 bottom right; © **John Gurche:** 124, 136–37, 156, 182, 204–5; **O.P. Hay** (from *Proceedings of the Washington Academy of Sciences,* XII, 1910): 100 top right; **G. Heilmann** (from *The Origin of Birds,* London, 1927): 115, 116, 196, 206, 239 bottom; **E. Hitchcock** (from *American Journal of Science,* 29, 1836): 67; **W. J. Holland** (from *American Nature,* 44, 1910): 98, 100 left and centre; **Humboldt University,** Natural History Museum, Berlin: 195; **Institut Royal des Sciences Naturelles de Belgique;** 64, 92;

Institute of Paleobiology, Polish Academy of Sciences: 108 (Bayn Dzak); Iain Jackson: 20–1; Arril Johnson: 138, 140, 190, 208 (all models at The City of Bristol Museum & Art Gallery); Martin Knowelden: 25; Martin Lockley: 186; Mansell Collection: 16 bottom; M L Design: 33, 35, 37, 38, 86, 146, 170 bottom, 223, 231, 232, 246; Steve Morris (models by Peter Minister): 211, 266, 272 bottom, 273 bottom; Museum of the Rockies: 184; Natural History Museum, London: 71 right, 82, 176, 194 right; David Nicholls: 123 (after John H. Ostrom), 164, 166; David Norman: 23, 28 all, 47, 65, 71 left, 94 top left, 120 left and top, 139, 242, 243, 244, 267, 272 top left, top right; H.F. Osborn (from Science, 10, 1899): 68 bottom; R. Owen (from Geology and Inhabitants of the Ancient World, London, 1854): 72; Oxford Scientific Films: 213 (Breck P. Kent); 225 (Hjalmar R. Bardarson); Oxford University Museum: 70; © Gregory S. Paul: 8, 46, 88; Peabody Museum of Natural History, Yale University: 80; Peter Phillips: 235 (model by Lyons Model Effects Ltd), 239 top right; Robin E.H. Reid: 173 all; Andrew Robinson: 52–3, 54–5, 56, 58–9, 176–7 all; Royal Ontario Museum, Toronto, Canada: 120 bottom right; John A. Ryder from Osborn, H.F. and Mook, C.C., 1919. Characters and Restoration of the Sauropod Genus Camarasaurus Cope, LVIII, no. 6, p.392: 96–7; Science Photo Library: 22 (Martin Dohrn); 222 (Jan Hinsch); © John Sibbick: front cover 2–3, 10, 39, 40–1, 42–3, 44–5, 101, 108 bottom, 110 right, 111, 162–3, 167, 169, 170, 180–1 colour, 234, 240, 254 left, 256, 257, 276 left; The Telegraph Colour Library: 210, 219; John Wilson: 112.

Publishers' Acknowledgments

The publishers would like to thank the following for their help, support and advice:
Jane Armstrong; Rod Caird; Allison Denyer; Ruth Goldwyn; Ralph Hancock; Kit Johnson; David Johnson; Peter Phillips; Angus Ross; Sue Scott; Bridgette Sibbick; John Sibbick; Merilyn Thorold; Viv Williams.

Dinosaur! series

Executive producers: Rod Caird, W. Paterson Ferns, Alfred Payrleitner, Michael von Wolkenstein; Series editor: Allan Segal; Directors: Jim Black, Christopher Rowley; Production manager: Allison Denyer; Series consultant: Dr David Norman; Researchers: Jane Armstrong, Ruth Goldwyn; Graphic designer: Peter Phillips; Model makers: Peter Minister, Model FX and Rachel Nettles, Lyons Model Effects Ltd; Illustrator: John Sibbick

INDEX

*Numbers in **bold** refer to illustrations*

288